Lab Workbook
Modern Plumbing

by

Charles H. Owenby

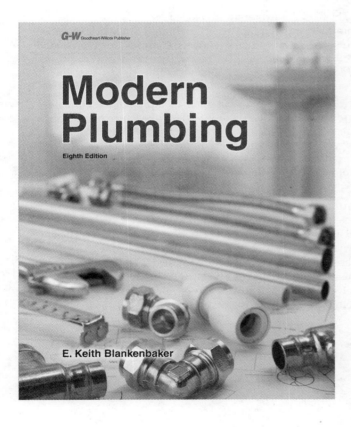

Publisher
The Goodheart-Willcox Company, Inc.
Tinley Park, IL
www.g-w.com

Copyright © 2015
by
The Goodheart-Willcox Company, Inc.

Manufactured in the United States of America.

ISBN 978-1-61960-868-9

7 8 9 – 15 – 22 21 20

The Goodheart-Willcox Company, Inc. Brand Disclaimer: Brand names, company names, and illustrations for products and services included in this text are provided for educational purposes only and do not represent or imply endorsement or recommendation by the author or the publisher.

The Goodheart-Willcox Company, Inc. Safety Notice: The reader is expressly advised to carefully read, understand, and apply all safety precautions and warnings described in this book or that might also be indicated in undertaking the activities and exercises described herein to minimize risk of personal injury or injury to others. Common sense and good judgment should also be exercised and applied to help avoid all potential hazards. The reader should always refer to the appropriate manufacturer's technical information, directions, and recommendations; then proceed with care to follow specific equipment operating instructions. The reader should understand these notices and cautions are not exhaustive.

The publisher makes no warranty or representation whatsoever, either expressed or implied, including but not limited to equipment, procedures, and applications described or referred to herein, their quality, performance, merchantability, or fitness for a particular purpose. The publisher assumes no responsibility for any changes, errors, or omissions in this book. The publisher specifically disclaims any liability whatsoever, including any direct, indirect, incidental, consequential, special, or exemplary damages resulting, in whole or in part, from the reader's use or reliance upon the information, instructions, procedures, warnings, cautions, applications, or other matter contained in this book. The publisher assumes no responsibility for the activities of the reader.

The Goodheart-Willcox Company, Inc. Internet Disclaimer: The Internet resources and listings in this Goodheart-Willcox Publisher product are provided solely as a convenience to you. These resources and listings were reviewed at the time of publication to provide you with accurate, safe, and appropriate information. Goodheart-Willcox Publisher has no control over the referenced websites and, due to the dynamic nature of the Internet, is not responsible or liable for the content, products, or performance of links to other websites or resources. Goodheart-Willcox Publisher makes no representation, either expressed or implied, regarding the content of these websites, and such references do not constitute an endorsement or recommendation of the information or content presented. It is your responsibility to take all protective measures to guard against inappropriate content, viruses, or other destructive elements.

Introduction

This lab workbook is designed to be used with the *Modern Plumbing* textbook. The chapter review questions and the jobs are aimed at helping you master the subject matter provided in the textbook. The questions will help you remember important ideas, theories, and concepts. The jobs will incorporate these ideas to help you improve your hands-on techniques and skills. Each chapter review corresponds to a chapter in the textbook, while a job may incorporate the ideas from several chapters of the textbook.

After studying a chapter in the textbook, complete the Test Your Knowledge questions and the Suggested Activities. Once these are finished, complete the corresponding Chapter Review. After studying each chapter and successfully completing the Chapter Review, complete the job(s) that relate to the chapter(s) you are studying.

Chapter Reviews

Chapter Reviews are to be used after studying a chapter in the textbook. Answer as many of the questions as possible without using your textbook. Complete the remaining questions by referring to the textbook where the topic is discussed. These Chapter Reviews can also be used as pretests to determine your knowledge level prior to studying a chapter.

Jobs

Jobs are to be completed after studying the textbook chapter(s) and the corresponding Chapter Review(s). The jobs allow you to apply the ideas, theories, and concepts learned in the textbook. The jobs will incorporate these ideas to help you improve your hands-on techniques and skills.

Jobs may consist of an Introduction, Tools and Equipment section, Textbook References, Objective, and Instructions. The Introduction is general information about the need and purpose of the job. The Tools and Equipment section is a detailed list of the tools and equipment needed to complete the job. *Gather all the tools and equipment on the list before starting the job.* The Textbook References are a list of textbook chapter(s) and pages that are to be used as reference for the job. The Objective is a detailed paragraph that describes the purpose and the goal of the job. The Instructions are a numbered, step-by-step procedure for completing the job. Each step is followed by a box with the word "Completed". Upon completion of each step, place a check in the box. Throughout the jobs there are highlighted notes that give you additional information about the job or step. Also throughout the jobs, there are highlighted cautions and warnings that give you additional safety information.

After completing the job, have your instructor evaluate your work and initial and date the job sheet. You will be evaluated on the final product, following the procedural steps, tool and equipment selection, return of tools and equipment, cleanliness of work area, and general shop safety. *Although time is a factor, **safety** and **work quality** are not to be sacrificed for speed.*

Safety

Safety is the number one priority when participating in hands-on jobs or when doing any plumbing work. The work area, or shop, is a place to work. It is not a place for "horseplay" or joke playing.

Your work area should always be clean and clear of debris. Make sure the proper tools and equipment are selected to complete the job. When your work is complete, make sure that all tools and equipment are in working order and are placed back in their proper place. Make sure the work area is clean when you have completed the job.

Always protect yourself. Wear safety glasses, goggles, or a face shield when working in the shop. Dress properly. Wear protective gloves, and avoid wearing loose-fitting clothing. Long sleeves should be rolled up, and if you wear a tie, remove it. Sturdy shoes with thick soles should be worn. Shoes with leather soles should be avoided. Leather soles have a tendency to have less traction. Rings, watches, and other jewelry should always be removed.

Always handle sharp and/or pointed objects with care. Tools such as scribers, cutting tools, and screwdrivers should be placed down with the cutting edges or point facing away from you. Avoid carrying any tool in your pocket.

When using solvents and adhesives, always read the instructions carefully. Some of these fluids require a well-ventilated work area because of the fumes that they emit. Some can also cause skin irritation if they come into contact with your skin. These fluids may also be flammable. Approved fire prevention practices should be followed in the work area while working with such fluids.

When using a torch or a melting furnace, substantial heat can be produced. Care must be taken when handling heated materials. When handling or moving with heated materials, make sure your walking path is clear. If you are handling a heated liquid and it is to be set down, make sure it is set down on a level surface.

Knowledge of first aid is very important. You should know and understand common first aid procedures. You should know where to locate the first aid kit in the shop.

Your safety and the safety of those around you is the highest priority. Think first before you perform any job or procedural step. If you are not sure about how to perform the procedure, ask your instructor for further assistance.

These jobs are intended to give you the opportunity to improve your hands-on skills as they pertain to the plumbing field. Do not attempt these jobs unless you are under the supervision of a trained plumbing instructor.

Table of Contents

Jobs   Page

Safety

Objective: You will be able to develop a list of general safety rules relating to clothing, ladders, electrical tools, and scaffolds and explain why it is necessary to develop safe working habits.

Instructions: *Carefully read Chapter 1 in the text and answer the following questions.*

_____ 1. Safety is one aspect of plumbing that is often neglected.
   A. True
   B. False

_____ 2. _____ contributes to accidents on the job.
   A. Carelessness
   B. Lack of knowledge
   C. Substance abuse
   D. All of the above.

_____ 3. Good housekeeping means keeping the work area as clean and orderly as possible.
   A. True
   B. False

_____ 4. Safe work habits are a result of attitudes formed by the worker.
   A. True
   B. False

_____ 5. Proper clothing can reduce the possibility of an accident or injury.
   A. True
   B. False

_____ 6. The National Safety Council estimates that _____% of the eye injuries that occur annually in the United States could be prevented by wearing protective eyewear.
   A. 50
   B. 65
   C. 75
   D. 90

_____ 7. If you wear corrective lenses, they must meet the _____ standard for safety glasses.

_____ 8. Corrective lenses worn on the job should be fitted with _____.

_____   9. Because it is more comfortable, loose clothing should be worn
                         while working with revolving machinery parts.
                         A. True
                         B. False

_____   10. Pants with cuffs should be worn on the job.
                         A. True
                         B. False

_____   11. Lifting or carrying materials incorrectly is most likely to cause
                         serious injury to your _____.
                         A. back
                         B. legs
                         C. arms
                         D. All of the above.

_____   12. When lifting, keep your back _____ and use your _____ muscles to
_____   raise the object.

_____   13. When lifting, repeated minor strains may result in permanent
                         damage.
                         A. True
                         B. False

_____   14. When carrying large pieces of pipe, use _____ to make the job
                         easier and safer.

_____   15. Most ladder-related injuries are caused by using the wrong ladder
_____   for the job, or by _____ or _____ climbing equipment.

_____   16. Stepladders are often used for indoor work.
                         A. True
                         B. False

_____   17. The duty rating of the ladder is its maximum _____ capacity.

_____   18. An all aluminum ladder should *not* be used near electrical wire
                         because it will _____ electricity.

_____   19. Extension ladders should be _____ to prevent them from slipping.
                         A. secured or held by a fellow worker
                         B. placed at 75° angle to the ground
                         C. placed at 80° angle to the ground
                         D. Both A and B.

_____   20. When using a stepladder, never work higher than the _____ rung
                         from the top of the ladder.

_____   21. Scaffolds 4' to 10' high and less than _____" wide must also be
                         guarded with rails.

_____   22. Never place a scaffold plank board on the top of guardrails.
                         A. True
                         B. False

Name _____

_____ 23. Electrical tools should be properly _____.
A. painted
B. grounded
C. lubricated
D. Both A and C.

_____ 24. The letters GFCI stand for _____.
A. Gas and Fire Code Institute
B. gas fired combustion ignition
C. ground fault circuit interrupter
D. None of the above.

_____ 25. Tool guards may be removed for the convenience of the worker.
A. True
B. False

_____ 26. _____ are potential fire hazards.
A. Compressed gas cylinders
B. Solvents
C. Liquid fuel
D. All of the above.

27. Match each of the following classes of fire with its description.

_____ a. Fires in combustible metals such as sodium, magnesium, and titanium. These fires require the use of extinguishing agents that will not react with the burning metals. Also, the extinguishing agent needs to be heat absorbing.

A. Class A fires
B. Class B fires
C. Class C fires
D. Class D fires

_____ b. Fires in "hot" electrical equipment are especially dangerous because only nonconductive extinguishing agents can be safely used. Note that once the electricity has been turned off, Class A or Class B extinguishers may be safe to use.

_____ c. Fires in flammable liquids such as gasoline, grease, and oil. Preventing oxygen (air) from mixing with the vapors from the flammable liquid results in smothering the fire.

_____ d. Fires in ordinary combustible materials such as paper, wood, cloth, rubber, and many plastics. The cooling effects of water or solutions of water and chemicals will extinguish the fire. Also, the coating effects of selected dry chemicals that retard combustion can control these fires.

_____ 28. Excavating and trenching work involves three major types of safety. They are protecting _____, workers in trenches, and using hand tools safely.

_____ 29. The two categories of fall protection are _____ and _____.
_____

_____ 30. Hand tool safety is a matter of _____.
   A. common standards
   B. common sense
   C. common knowledge
   D. common sizes

_____ 31. Mushroomed heads on chisels and similar tools must *not* be used.
   A. True
   B. False

_____ 32. Many materials used on a construction site are health hazards.
   A. True
   B. False

_____ 33. The federal _____ Standard was adopted to provide information to
   workers about the materials they are working with and the potential hazards these materials present.

_____ 34. Employers are required to make information available to
   employees about both physical and _____ hazards.

_____ 35. Producers of hazardous materials must label all containers of
   hazardous materials to identify the chemicals and provide appropriate warnings.
   A. True
   B. False

_____ 36. A(n) _____ contains detailed information on each chemical in a
   specific product.

37. Identify the following acronyms.

   a. TLV _____

   b. PEL_____

   c. TWA_____

   e. STEL _____

   e. SDS_____

38. List four ways hazardous chemicals or materials can enter the body.

   _____

   _____

   _____

   _____

_____ 39. A(n) _____ will protect the worker from inhaling hazardous
   chemicals.

Name _____

_____ 40. One respirator will work for all job site hazardous conditions.
  A. True
  B. False

_____ 41. Vials of _____ oil are used to test the fit of a respirator.
  A. apple
  B. banana
  C. orange
  D. peach

_____ 42. Full-face respirators provide _____ protection and are less likely to
_____ leak at the joint between the filter and _____.

_____ 43. Respirators should be inspected for damage or wear before and
after each use.
  A. True
  B. False

_____ 44. Hearing loss occurs only when workers are exposed to very loud
sounds.
  A. True
  B. False

_____ 45. According to NIOSH, you should *not* enter an area where the
oxygen level is below _____.
  A. 18%
  B. 18.5%
  C. 19%
  D. 19.5%

46. What effect does an oxygen-deficient atmosphere have on the body at the following levels?

  a. 19.5% _____

  b. 16% _____

  c. 14% _____

  d. 6% _____

_____ 47. In a "confined space," carbon monoxide can be generated when
materials are brazed or soldered.
  A. True
  B. False

_____ 48. Working safely in a confined space requires forced _____ at a
minimum.

49. What are the two most common bloodborne pathogens?

  _____

  _____

Name _____ Date_____

Period_____ Course _____

Score_____

Plumbing Tools

2

Objective: You will be able to identify the various plumbing tools and their uses. You will be able to select the proper tool for the desired task and explain how to maintain common plumbing tools.

Instructions: *Carefully read Chapter 2 in the text and answer the following questions.*

_____ 1. Measuring tools are used to measure _____.
    A. length, height, and diameter
    B. levelness
    C. plumbness
    D. All of the above.

_____ 2. Layout tools are used to produce accurate _____, _____, or any
_____   other marking.

_____ 3. Common lengths of long steel measuring tapes are 25′, 50′, 100′,
    200′, and 300′ lengths.
    A. True
    B. False

_____ 4. Plumber's rules are special folding rules available in _____′ and
_____   _____′ lengths.

_____ 5. Combination squares can measure _____° angles.
    A. 90
    B. 45
    C. 60
    D. Both A and B.

  6. List three marking tools used to mark various types of pipe.

_____

_____

_____ 7. Plumb is to vertical, as level is to _____.
    A. straight
    B. slope
    C. horizontal
    D. parallel

_____   8. Chalk lines are used to lay out long, straight lines on hard and rather smooth surfaces.
   A. True
   B. False

_____   9. A(n) _____ is an instrument that has a pencil in one leg and is used to lay out circles and arcs.

10. Match the toothed cutting tools with their proper names. Place the appropriate letter in the blank.

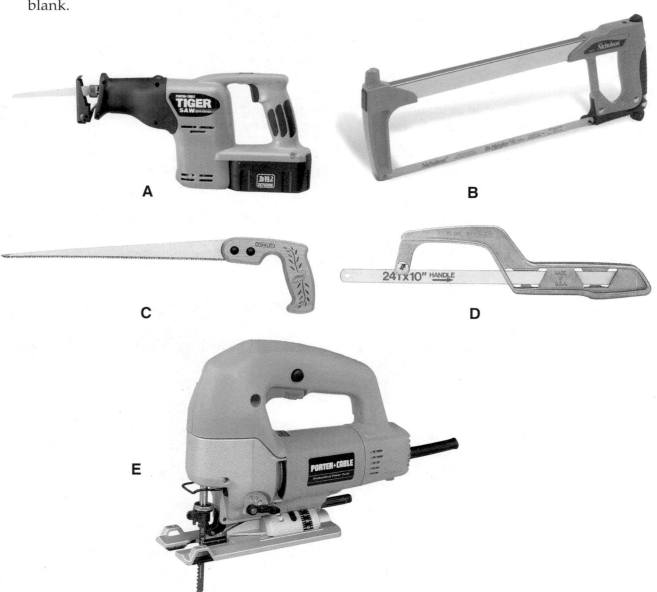

Porter-Cable; Cooper Tools; Stanley Tools

_____ a. Hacksaw

_____ b. Compass saw

_____ c. Reciprocating saw

_____ d. Jab saw

_____ e. Saber saw

Name _____

_____ 11. Files are toothed cutting tools that have one common cross-sectional shape.
　A. True
　B. False

_____ 12. For best results, the wood chisel is ground to a _____° angle.
　A. 10
　B. 25
　C. 30
　D. 45

_____ 13. For best results, the cold chisel is ground to a _____° angle.
　A. 22.5
　B. 25
　C. 30
　D. 60

14. Match each of the following with its proper name. Place the appropriate letter in the blank.

Makita USA, Inc.; Milwaukee Electric Tool Corp.; Greenlee Textron; mayer kleinostheim/Shutterstock.com; Goodheart-Willcox Publisher

_____ a. Multispur bit

_____ b. Offset portable drill

_____ c. Hole saw

_____ d. Plumbers' auger

_____ e. Spade bit

_____ 15. Plumber's augers are available in diameters of _____" through
_____".
   A. 1/4; 1
   B. 3/4; 1-1/2
   C. 1; 2
   D. 1-1/2; 3

_____ 16. Multispur bits, plumber's augers, and _____ bits are tools used for
drilling holes.

_____ 17. The operation of _____ removes the burr that is formed inside a
pipe during cutting.

_____ 18. A _____ threading die is a special die used for threading galva-
nized steel pipe.
   A. special
   B. bolt
   C. pipe
   D. straight

_____ 19. _____ are sometimes needed to hold plumbing parts while the
operations are performed on them.
   A. Drills
   B. Vises
   C. Pipes
   D. Reamers

_____ 20. To protect pipe dies when cutting threads, _____ is applied to the
threads.
   A. machine oil
   B. cutting oil
   C. 30 weight motor oil
   D. water

_____ 21. The three-way pipe die and stock permits three diameters of pipe
to be threaded with a single tool.
   A. True
   B. False

_____ 22. A pipe wrench is used to _____ threaded pipe during assembly.
   A. hold or turn
   B. adjust
   C. ream
   D. None of the above.

23. Pipe wrenches are available in what three basic designs?

_____

_____ 24. A _____ wrench is used to assemble chrome-plated pipe so
damage to the pipe does not occur.
   A. strap
   B. chain
   C. pipe
   D. None of the above.

Name _____

25. Match each of the wrenches with its proper name.

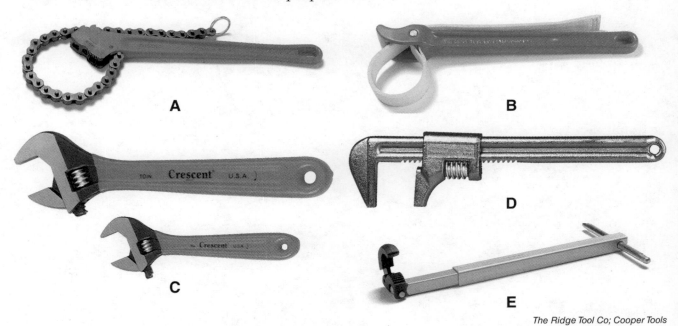

The Ridge Tool Co; Cooper Tools

_____ a. Monkey wrench

_____ b. Chain wrench

_____ c. Strap wrench

_____ d. Basin wrench

_____ e. Adjustable wrench

_____ 26. A(n) _____ vise is the most common holding device used by plumbers.

_____ 27. A _____ hammer is used for driving or pulling nails and for tapping a wood chisel.
   A. ball peen
   B. sledge
   C. carpenter's
   D. None of the above.

_____ 28. A(n) _____ screwdriver generally has two straight and two Phillips blades.

Leveling Instruments

3

Objective: You will be able to explain the operation of the builder's level and laser level and describe how they are used to find levels and properly slope drainage pipe.

Instructions: *Carefully read Chapter 3 in the text and answer the following questions.*

_____ 1. Plumbers may use a _____ for leveling.
    A. level
    B. straightedge
    C. chalk line
    D. All of the above.

_____ 2. A builder's level is also known as a _____.
    A. surveyor's level
    B. transit
    C. line level
    D. Both A and B.

_____ 3. A builder's level mounts on a _____.
    A. bipod
    B. tripod
    C. bench
    D. pipe vise

_____ 4. A builder's level rotates _____ degrees.

_____ 5. A _____ is a graduated stick used to find elevations.
    A. pole rod
    B. stadia rod
    C. tape
    D. framing square

6. Match the parts of the builder's level with the proper part name.

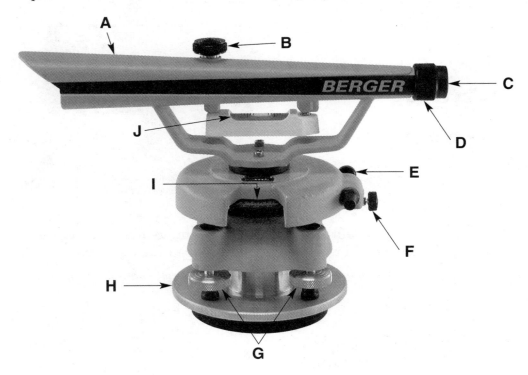

CST/berger

_____ a. Telescope

_____ b. Leveling screws

_____ c. Eyepiece

_____ d. Telescope focus knob

_____ e. Eyepiece focus

_____ f. Horizontal motion screw

_____ g. Telescope level

_____ h. Horizontal motion tangent screw

_____ i. Horizontal circle vernier

_____ j. Base plate

_____ 7. Using a builder's level, a plumber has a first reading of 6'-0", a second reading of 9'-4", and a length of run of 100'. Compute the rate of fall.

_____ 8. The horizontal crosshair in the telescope sight indicates the point at which the reading should be made.
A. True
B. False

_____ 9. A(n) _____ level is self-adjusting to maintain a horizontal line of site.

Name _____

_____ 10. A(n) _____ is an instrument that amplifies, or strengthens, light projecting it as a thin beam.

_____ 11. Long-term exposure to the laser light beam may cause eye injury.
A. True
B. False

_____ 12. A worker does *not* need to be qualified to operate laser equipment.
A. True
B. False

_____ 13. When a laser is being operated, beam shutters and caps should be used.
A. True
B. False

_____ 14. When a laser is left unattended, it should be _____.
A. turned off
B. recharged
C. left on
D. None of the above.

_____ 15. During installation of a sewer pipe using a laser, the laser beam is projected through the pipe until it strikes a(n) _____.
A. block
B. fitting
C. end cap
D. beam target

Name _____ Date _____

Period_____ Course _____

Score_____

Mathematics for Plumbers

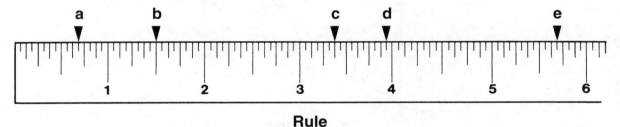

4

Objective: You will be able to read a rule, add and subtract fractions and whole numbers, find area and volume, and convert measurements.

Instructions: *Carefully read Chapter 4 in the text and answer the following questions.*

_____ 1. The _____ is the basic measuring tool used by plumbers.
    A. yardstick
    B. folding rule
    C. steel tape
    D. Both B and C.

2. Using the rule, identify the measurements.

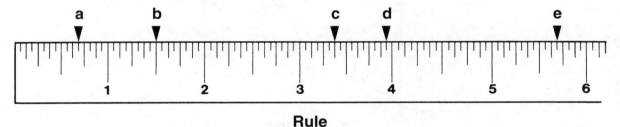

Rule

Goodheart-Willcox Publisher

a. _____

b. _____

c. _____

d. _____

e. _____

_____ 3. To prevent costly errors when cutting materials, measurements should be rechecked.
    A. True
    B. False

_____ 4. There are _____ 1/16's in 7/8.

_____ 5. There are _____ 1/8's in 3/4.

6. Solve the following problems.

_____   a. 14-1/2 + 2-1/2 = _____

_____   b. 22-3/4 − 3-1/4 = _____

_____   c. 8-5/8 + 2-3/16 = _____

_____   d. 10-3/4 + 1-3/8 = _____

_____   e. 12-1/2 − 2-3/4 = _____

_____ 7. A measurement of 12′ converts to _____″.

_____ 8. A measurement of 197′ converts to _____″.

_____ 9. In the fraction 1/2, the number 2 is the numerator.
        A. True
        B. False

_____ 10. Using pipe layout I, the distance of AB is _____″.

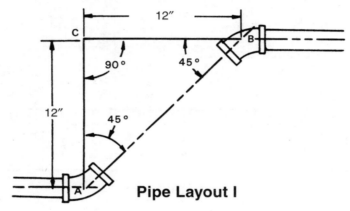

Pipe Layout I

_____ 11. Using pipe layout II, the distance of AB is _____″.

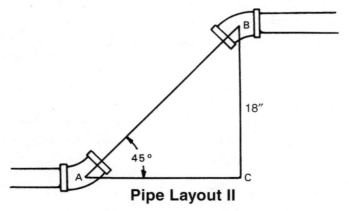

Pipe Layout II

Name _____

_____ 12. Using pipe layout III, the distance of AC is _____".

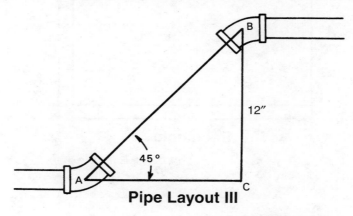

Pipe Layout III

13. Using pipe layout IV, identify the run, offset, and travel.

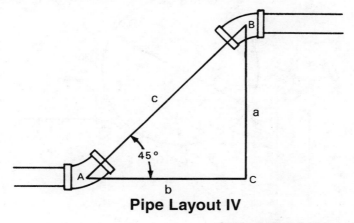

Pipe Layout IV

a. _____

b. _____

c. _____

_____ 14. The surface area of a square or a rectangular surface can be computed by:
A. multiplying the length, width, and height.
B. adding the length and the width.
C. multiplying the length times the width.
D. dividing the length by the width.

_____ 15. The area of the rectangle is _____ square feet.

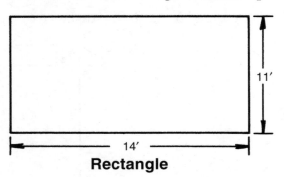

Rectangle

Goodheart-Willcox Publisher

_____ 16. The area of the circle is _____ square inches.

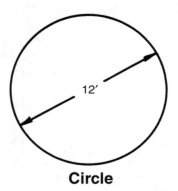

Circle

Goodheart-Willcox Publisher

_____ 17. The volume of the rectangular tank is _____ cubic feet.

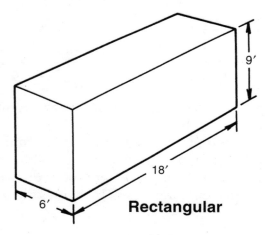

Rectangular

Goodheart-Willcox Publisher

Name _____

_____ 18. The volume of the cylindrical tank is _____ cubic feet.

Cylindrical Tank

Goodheart-Willcox Publisher

_____ 19. If 7.48 gallons of water equals one cubic foot, then _____ gallons of water can be held in the cylindrical tank.

_____ 20. To convert cubic inches to gallons, divide by _____.
A. 231
B. 1728
C. 7.48
D. 28.32

_____ 21. Six gallons equals _____ liters.
A. 3.78
B. 7.48
C. 22.68
D. 11.34

_____ 22. There are _____ meters in one mile.
A. 5280
B. 1609.344
C. 10250
D. 1333.225

_____ 23. To change from one metric unit to another, multiply or divide by multiples of 2.
A. True
B. False

_____ 24. The metric unit used for long distances is the _____.

_____ 25. The metric unit used for very small measurements is the _____.

_____ 26. The metric unit used for dry volume is the cubic centimeter.
A. True
B. False

27. Convert the following measurements.

_____ a. 16″ = _____ millimeter(s)

_____ b. 12″ = _____ meter(s)

_____ c. 9″ = _____ centimeter(s)

_____ d. 478 cubic inches = _____ liter(s)

_____ e. 3.75 cubic feet = _____ liter(s)

_____ f. 24 cubic inches = _____ cubic feet

_____ g. 20 cubic feet = _____ cubic yard(s)

_____ h. 12 cubic inches = _____ gallon(s)

_____ i. 5 cubic feet = _____ gallon(s)

_____ j. 35 cubic feet = _____ cubic meter(s)

_____ 28. The newton is the metric unit for force.
    A. True
    B. False

_____ 29. _____ scale is used to measure temperatures in the supercold range of –273°C and up.

30. Convert the following temperatures.

_____ a. 20°C = _____ °K

_____ b. 220°F = _____ °C

_____ c. 25°C = _____ °F

_____ d. 298°K = _____ °C

_____ e. 72°F = _____ °C

Hydraulics and Pneumatics

5

Objective: You will be able to explain and apply characteristics of fluids under pressure, apply practical methods of computing useful pressures, and demonstrate undesirable characteristics of gases in incorrectly designed drainage systems.

Instructions: *Carefully read Chapter 5 in the text and answer the following questions.*

_____ 1. _____ is the study of the characteristics of fluids.

_____ 2. A cubic foot of water weighs _____ pounds.
   A. 14.8
   B. 62.4
   C. 82.4
   D. 100

_____ 3. One cubic foot of water exerts _____ pounds of pressure per square foot.

_____ 4. Water pressure decreases as the depth of water increases.
   A. True
   B. False

_____ 5. If water is stored in a tower 175 feet tall, the water pressure at the base of the tower is _____ psi.

_____ 6. _____ is the pressure available at some point in the water system, and is measured by the depth of a column of water above the point where the measurement is taken.

_____ 7. For scientific studies, it is known that a column of water 1' high produces a pressure of _____ psi.
   A. 0.43
   B. 0.86
   C. 1.20
   D. 1.40

_____ 8. Plumbers must consider the effects of friction on water pressure and flow rates.
   A. True
   B. False

_____ 9. Streamlined flow produces considerable friction, while turbulent flow produces little friction.
A. True
B. False

_____ 10. Compared to schedule 40 plastic pipe, grade L copper pipe has _____ (more/less) head loss.

_____ 11. Water hammer can cause pipes to vibrate and possibly burst.
A. True
B. False

_____ 12. _____ is the study of compressible gases.

_____ 13. A(n) _____ bend in a low-pressure piping system is likely to cause an air lock that blocks the flow through the pipe.

_____ 14. Under normal conditions, both the sanitary and storm sewer are completely filled with water.
A. True
B. False

_____ 15. In the following illustration, the DWV piping system shown for Building B is better than the system in Building A.
A. True
B. False

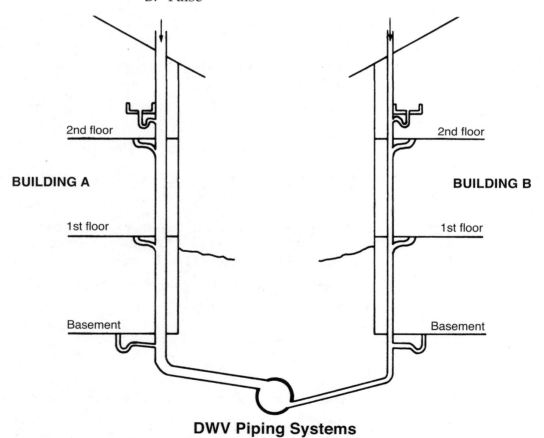

DWV Piping Systems

Goodheart-Willcox Publisher

Print Reading and Sketching

6

Objective: You will be able to identify the plumbing symbols and abbreviations, interpret various kinds of plans, and take dimensions of drawings in inches and feet. You will be able to scale drawings and prepare two- and three-dimensional piping sketches.

Instructions: *Carefully read Chapter 6 in the text and answer the following questions.*

_____ 1. Scale drawings may use fractions of inches to represent feet.
    A. True
    B. False

_____ 2. Drawings are frequently referred to as _____.
    A. prints
    B. blueprints
    C. master plans
    D. Both A and B.

_____ 3. A _____ may use prints.
    A. building official
    B. contractor
    C. supplier
    D. All of the above.

4. Match the plan description with the proper plan.

	Plan	**Plan Description**
_____	a. Plot plan	A. Construction of walls, stairs, and doors
_____	b. Elevation drawings	B. Describes arrangement of rooms
_____	c. Floor plan	C. Describes size and shape of footing
_____	d. Foundation plan	D. Location of structure on the lot
_____	e. Detail drawings	E. Exterior appearance

_____   5. _____ are a set of instructions that provide information about materials and work quality.
   A. Blueprints
   B. Specifications
   C. Catalogs
   D. None of the above.

_____   6. A section view indicates the _____ distances critical to the installation of DWV piping.

_____   7. Dimensions on architectural drawings are usually given in _____.
   A. meters
   B. feet and inches
   C. inches only
   D. feet only

_____   8. A distance of 83 inches is commonly written as _____.
   A. 83'
   B. 6'-11"
   C. 5'-11"
   D. None of the above.

_____   9. _____ are symbols used to indicate the limits of a particular dimension.
   A. Arrowheads
   B. Dots
   C. Diagonal lines
   D. All of the above.

10. Find, in the simplest form, the total dimension of 6'-10", 3'-3", and 8'-11"

_____ a. Total feet

_____ b. Total inches

_____ c. Simplified

11. Match the symbols with plumbing fixtures, appliances, and mechanical equipment.

_____ a. Dry well

_____ b. Range

_____ c. Sump pit

_____ d. Water heater

_____ e. Vacuum outlet

_____ f. Hose bib

_____ g. Floor drain

_____ h. Dishwasher

_____ i. Built-in cooking top

_____ j. Water softener

Symbols

Name _____

12. Identify the abbreviations for the following.

		Description	Abbreviation

_____ a. Cast iron	A. FD
_____ b. Clean out	B. HW
_____ c. Cold water	C. CI
_____ d. Hot water	D. CO
_____ e. Dishwasher	E. HB
_____ f. Floor drain	F. GI
_____ g. Hose bibb	G. PLBG
_____ h. Galvanized iron	H. LT
_____ i. Plastic	I. PLAS
_____ j. Lavatory	J. WC
	K. CL
	M. LAV
	N. WS
	O. CW

_____ 13. The following symbol represents a _____ connection.

Goodheart-Willcox Publisher

A. soldered or cemented
B. bell and spigot
C. screwed

_____ 14. The following symbol represents a(n) _____.

Goodheart-Willcox Publisher

A. Reducing elbow
B. Elbow—90°
C. Elbow—45°

_____ 15. To find the correct scale of a drawing, the title block is referenced.
A. True
B. False

_____ 16. Building plans are customarily drawn to a _____ scale.
  A. 1/2 inch = 1 foot
  B. 1/4 inch = 1 foot
  C. 1 mm = 50 mm
  D. Both B and C.

_____ 17. A recessed tub has a wall on three sides.
  A. True
  B. False

18. Find the dimension of each opening.

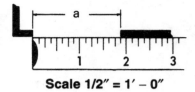

Scale 1/2″ = 1′ – 0″

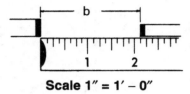

Scale 1″ = 1′ – 0″

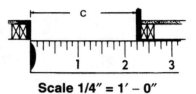

Scale 1/4″ = 1′ – 0″

Goodheart-Willcox Publisher

a. _____

b. _____

c. _____

19. What are the three types of piping sketches?

_____

_____

_____

Name _____

_____ 20. An isometric sketch is the simplest type of piping sketch.
- A. True
- B. False

_____ 21. The _____ is a piping sketch that is frequently drawn at the job site.

22. Using 1, 2, 3, and 4, indicate the order of the step in making a plan view sketch.

_____ a. Darken lines and add notes

_____ b. Lightly draw pipe

_____ c. Locate known points

_____ d. Indicate fitting symbols

_____ 23. In an isometric axis, converging lines form angles of _____° or _____° from horizontal.
- A. 90; 30
- B. 90; 45
- C. 120; 30
- D. 120; 45

_____ 24. Isometric sketches are helpful in illustrating more complex piping systems.
- A. True
- B. False

_____ 25. _____ grid paper is extremely helpful in making an isometric sketch.
- A. Horizontal
- B. Vertical
- C. Isometric
- D. None of the above.

Rigging and Hoisting

7

Objective: You will be able to describe ropes and other tools used for hoisting, demonstrate methods of securing rope to piping, and describe the types of ladders.

Instructions: *Carefully read Chapter 7 in the text and answer the following questions.*

_____ 1. Natural fiber ropes are used extensively for construction work.
  A. True
  B. False

_____ 2. _____ is made from the fiber of a wild banana plant.

3. Match each type of rope with its characteristic.

Rope Type	Characteristic
_____ a. American hemp rope	A. 80% as strong as No. 1 manila rope
_____ b. Sisal rope	B. 60% as strong as No. 1 manila rope
_____ c. Cotton rope	C. 50% as strong as No. 1 manila rope

_____ 4. _____ is a type of manufactured fibers from which some ropes are made.
  A. Nylon
  B. Rayon
  C. Glass
  D. All of the above.

_____ 5. Hemp fibers are stronger than Dacron® fibers.
  A. True
  B. False

_____ 6. Synthetic fiber ropes resist rot and deterioration.
  A. True
  B. False

_____ 7. Because of possible unwanted wear, ropes should not be _____.
  A. kept dry
  B. run across sharp surfaces
  C. frozen
  D. Both B and C.

8. Match each knot with its proper name.

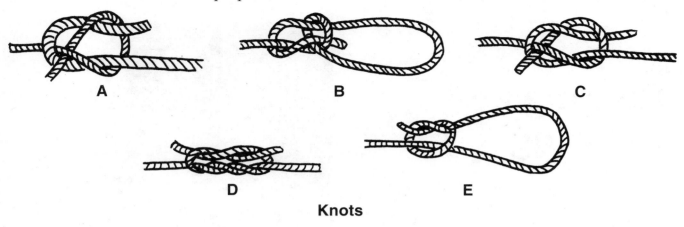

Knots

Goodheart-Willcox Publisher

_____ a. Slip knot

_____ b. Bowline

_____ c. Sheet bend

_____ d. Single carrick bend

_____ e. Surgeons' knot

_____ 9. A _____ is attached to scaffold and permits loads to be lifted by
a person at the base of the scaffold.
A. lever hoist
B. hoist standard
C. half hitch chain
D. winch crane

_____ 10. A portable _____ is a hoisting device that is used in large buildings
to lift large pipes.

_____ 11. A _____ is a hoisting device that has a ratchet mechanism that
causes it to be operated slower than other devices.
A. lever hoist
B. hoist standard
C. winch crane
D. ratchet lift

_____ 12. A fresh coat of paint should be applied to a wooden ladder annually.
A. True
B. False

_____ 13. A(n) _____ ladder has two or more sections that can be extended
to adjust the length.

_____ 14. A(n) _____ is self-supporting, nonadjustable, and ranges from 4' to
16' in length.

_____ 15. A(n) _____ ladder has hinged joints, enabling it to be used in a
variety of configurations.

Building and Plumbing Codes

8

Objective: You will be able to explain zoning laws and building codes and how they are administered and enforced. You will be able to list points a plumbing code should cover and be able to apply code requirements to a plumbing installation.

Instructions: *Carefully read Chapter 8 in the text and answer the following questions.*

_____ 1. Minimum standards in buildings are specified to provide for the health and safety of the people.
   A. True
   B. False

_____ 2. _____ are laws that regulate the type of structure that can be built in a given area.
   A. Zoning laws
   B. Plumbing codes
   C. Electrical codes
   D. Building standards

_____ 3. Within these laws, residential, office, and industrial activities are *not* separated.
   A. True
   B. False

_____ 4. Building codes control the _____.
   A. quality of building materials
   B. quality of work
   C. number and type of exits
   D. All of the above.

_____ 5. Plumbing and electrical codes are almost always combined with the building code.
   A. True
   B. False

_____ 6. The enforcement of the plumbing code is generally delegated to the National Plumbing Commission.
   A. True
   B. False

_____ 7. When applying for a building permit, the contractor must submit two copies of _____ and _____ to the building inspector.
A. plans; regulations
B. codes; blueprints
C. plans; specifications
D. sepias; codes

_____ 8. After a permit is obtained, a construction project will be inspected only after its completion.
A. True
B. False

_____ 9. If work being done does not meet the codes, building officials can obtain a court order to stop construction.
A. True
B. False

_____ 10. To improve the quality of codes and provide a degree of standard-ization, _____ codes were developed.

_____ 11. _____ are involved in the development of plumbing codes.
A. Local governments
B. Architects
C. Plumbers
D. All of the above.

_____ 12. Before a model plumbing code can be enforced, it must be adopted by the local government.
A. True
B. False

13. Match the model code with its sponsoring organization.

	Model Code	Sponsoring Organization
_____ a.	BOCA Basic Building Code	A. Building Officials and Code Administrators International, Inc.
_____ b.	ICBO Plumbing Code	B. International Association of Plumbing and Mechanical Officials
_____ c.	One- and Two-Family Dwelling Code	C. Council of American Building Officials
_____ d.	Standard Plumbing Code	D. Southern Building Code Congress International, Inc.
_____ e.	Uniform Plumbing Code	E. International Conference of Building Officials

_____ 14. Plumbers have the authority to modify the plumbing code to meet their needs.
A. True
B. False

Name _____

15. List three types of information that should be contained in a well-written plumbing code.

_____

_____

_____

_____

_____

_____

Name _____ Date _____

Period _____ Course _____

Score _____

Soldering, Brazing, and Welding

Objective: You will be able to identify solders, brazing filler metals, and fluxes used for soldering and brazing copper pipe and fittings. You will be able to describe the processes by which pipe and fittings are joined.

Instructions: *Carefully read Chapter 9 in the text and answer the following questions.*

_____ 1. _____ is the method of using heat to form joints between two metallic surfaces using a nonferrous filler metal.
A. Flaring
B. Swagging
C. Soldering
D. Lasering

_____ 2. A nonferrous metal does *not* contain _____.
A. aluminum
B. iron
C. copper
D. flux

_____ 3. Capillary attraction is the amount of attraction a metal has to a magnet.
A. True
B. False

_____ 4. Solder composed of 50% tin and 50% lead should be used for potable water supply piping.
A. True
B. False

5. Hard solders are used in the brazing of what materials?

_____

_____

_____ 6. _____ is the process of picking up oxygen that produces tarnish and rust in metals.

_____ 7. _____ fluxes are best suited for plumbing work.
  A. Corrosive
  B. Noncorrosive
  C. Less corrosive
  D. None of the above.

_____ 8. It is important to clean the copper pipe and fittings before applying the flux.
  A. True
  B. False

9. Match the cleaning tools with their proper name.

A

Photo courtesy of Oatey SCS

B   **C**

Goodheart-Willcox Publisher

_____ a. Tube brush

_____ b. Fitting brush

_____ c. Reamer

_____ 10. When lighting a torch, _____.
  A. use a spark lighter
  B. direct tip away from you
  C. direct tip away from flammable materials
  D. All of the above.

_____ 11. The solder should be melted in the flame of the torch.
  A. True
  B. False

_____
_____ 12. Brazing is done at temperatures above _____°F (_____°C).

_____ 13. Brazing is the same process as braze welding.
  A. True
  B. False

_____ 14. The strength of a joint comes from the ability of the silver braze to flow into the porous grain structure of the base metal.
  A. True
  B. False

Name _____

_____ 15. To ensure an excellent bond when brazing, the clearance gap
between pipe and fitting must be only _____" to _____".
   A. .002; .004
   B. .002; .005
   C. .003; .004
   D. .003; .006

16. Match each of the braze materials with its proper description.

Braze Material	**Description**
_____ a. Nickel	A. Used for joining copper and copper alloy
_____ b. Copper-phosphorus	B. Used for brazing aluminum
_____ c. Silver	C. Used for extreme heat and corrosive resistance
_____ d. Copper and copper-zinc	D. Used for brazing all ferrous and nonferrous metals, except aluminum
_____ e. Aluminum-silicon	E. Used for brazing iron and steel

_____ 17. To braze 1-1/2" or 2" pipe, a No. 7 torch tip is recommended,
which require _____ psi oxygen pressure and _____ psi acetylene
pressure.

_____ 18. A neutral flame should *never* be used for brazing.
   A. True
   B. False

_____ 19. A carburizing flame is the result of an excess of _____.
   A. oxygen
   B. acetylene
   C. carbon
   D. Both A and B.

_____ 20. Welding in the plumbing industry is generally limited to repair
work on _____ pipe systems.

Excavating

10

Objective: Explain the importance of locating and protecting existing underground utilities. Explain the importance of protecting workers while trenching. Develop a list of general safety rules regarding the use of excavating tools and machines.

Instructions: *Carefully read Chapter 10 in the text and answer the following questions.*

_____ 1. All underground _____ must be located and protected before digging begins.

_____ 2. Excavating work associated with plumbing is done by _____.
A. backhoes
B. front loaders
C. trenchers
D. All of the above.

_____ 3. Trenchers or small backhoes can be used to produce narrow trenches.
A. True
B. False

_____ 4. _____ requires that excavations and trenches over _____' deep be guarded with shoring.
A. NFPA / 5
B. ASTM / 5
C. AWWA / 10
D. OSHA / 5

_____ 5. _____ may be used in place of shoring or sloping.
A. Sleeves
B. Trench boxes
C. Steel pipe
D. None of the above.

_____ 6. Vertical strut shoring is generally used in very unstable soils.
A. True
B. False

7. List two tools that can be used to loosen ground that is too hard to be loosened with a shovel.

_____

_____

_____ 8. When you complete the swing of a pick/mattock, the point of the tool will be _____ to the surface being struck.

_____ 9. When striking dense material with a pick/mattock, you and others around you must wear eye protection equipment.
A. True
B. False

_____ 10. Removing large quantities of dense material is best accomplished with a chipping _____.
A. compressor
B. chisel
C. hammer
D. pick

_____ 11. Pneumatic breakers or chipping hammers are commonly called _____.
A. jackhammers
B. rotary hammers
C. ball-peen hammers
D. sledgehammers

_____ 12. A _____ is used to provide a high volume of compressed air at the work site for pneumatic tools.
A. portable motor
B. portable compressor
C. portable hammer
D. chipping hammer

_____ 13. To prevent freezing of the water service pipe, the pipe must be installed below the _____ line.
A. fall
B. slope
C. frost
D. sidewalk

_____ 14. _____ can be placed in the bottom of a trench to support the pipe at the required slope.
A. Gravel
B. Cement
C. Sand
D. None of the above.

Name _____ Date _____

Period _____ Course _____

Score _____

Water Supply Systems

Objective: You will be able to list the types of wells, describe the various types of water pumps, and explain the operation of a pressure tank.

Instructions: *Carefully read Chapter 11 in the text and answer the following questions.*

_____ 1. The _____ cycle makes it possible for people to repeatedly use water.
   A. rain
   B. water
   C. evaporation
   D. None of the above.

_____ 2. All groundwater is safe for human consumption.
   A. True
   B. False

_____ 3. The _____ is responsible for monitoring water sources so they meet standards.
   A. EPA
   B. Safe Standards Institute
   C. Potable Water Agency
   D. None of the above.

_____ 4. Sulfur may be added to groundwater to purify it.
   A. True
   B. False

_____ 5. A(n) _____ is a potential source of well contamination.
   A. underground fuel tank
   B. livestock feedlot
   C. waste disposal site
   D. All of the above.

6. List the four most common types of wells.

_____

_____

_____ 7. Dug wells are seldom more than 20′ deep.
         A. True
         B. False

_____ 8. Driven wells are made by forcing a(n) _____ into the ground.

_____ 9. The two most commonly used methods of drilling wells are the
_____   _____ and the _____ methods.

_____ 10. The vertical distance that the water is to be lifted is an important
         factor in water pump selection.
         A. True
         B. False

_____ 11. Water weighs _____ pounds per gallon.
         A. four
         B. six
         C. eight
         D. ten

_____ 12. When the peak demand varies greatly from the average demand, a
         small storage should be used.
         A. True
         B. False

_____ 13. For a typical plumbing fixture, a minimum water pressure of
         _____ pounds per square inch is required.

_____ 14. The total distance the water must be pumped, the number of
         fittings, and the speed at which the water is traveling through the
         pipe all contribute to _____.
         A. volume available
         B. friction loss
         C. static pressure
         D. available water source

_____ 15. The atmosphere exerts a pressure of _____ psi on the surface of
         water.
         A. 11
         B. 12.77
         C. 14.7
         D. 30.7

_____ 16. _____ pumps lift water by moving a piston back and forth in a
         cylinder.
         A. Centrifugal
         B. Jet
         C. Rotary
         D. None of the above.

_____ 17. To lift water, centrifugal pumps use a rapid spinning _____, which
         is a heavy disk mounted in the pump housing.

Name _____

_____ 18. Jet pumps are difficult to use because of the number of moving parts in the well that are vulnerable to deterioration.
A. True
B. False

_____ 19. Rotary pumps have a constant discharge rate and operate efficiently.
A. True
B. False

_____ 20. Wells deeper than _____' require a pumping device installed within a well casing.

_____ 21. Each stage of a submersible centrifugal pump decreases water pressure.
A. True
B. False

_____ 22. Local codes may require that pumps be installed by a licensed worker.
A. True
B. False

_____ 23. A well must be sanitized to overcome any unsanitary condition caused by the drilling operation.
A. True
B. False

_____ 24. To prevent galvanic corrosion, a _____ fitting should be installed between a submersible pump and the drop pipe if they are dissimilar metals.
A. galvanized
B. copper
C. dielectric
D. DWV

_____ 25. Hydropneumatic tanks contain _____ under pressure.
A. air
B. water
C. oil
D. Both A and B.

_____ 26. Which of the following statements is true about hydropneumatic tanks?
A. As water volume builds up in the tank, so does the pressure.
B. Hydropneumatic tanks are economical for large water supply systems.
C. Supercharge is the amount of water pressure in the tank.
D. Both A and B.

_____ 27. As a general rule for sizing pressure tanks, the usable capacity should be five times the pump capacity.
A. True
B. False

_____ 28. The outside of the well casing and the drilled hole must be filled
to prevent surface water from entering the well.
A. True
B. False

_____ 29. To prevent freezing, a(n) _____ must be installed where water
piping is kept below the frost line.

_____ 30. Using a drill and tapping machine, it is possible to tap a water
main without turning off the water.
A. True
B. False

Name _____ Date _____

Period _____ Course _____

Score _____

Water Treatment

Objective: You will be able to identify water contaminants, describe devices to treat water, and know which devices have an effect on each of the contaminants.

Instructions: *Carefully read Chapter 12 in the text and answer the following questions.*

_____ 1. Bacteria that are considered harmful to humans are called _____.

_____ 2. _____ of the water supply is the most widely used means of killing bacteria.

_____ 3. Heavy metals that may be found in water include _____.
A. tetrachloride and toluene
B. mercury and lead
C. selenium and nitrates
D. arsenic and magnesium

_____ 4. A reddish stain on fixtures is produced by _____.
A. turbidity
B. high pH water
C. iron oxide
D. manganese

_____ 5. Plumbers are called on to install and maintain water treatment equipment.
A. True
B. False

_____ 6. Water that is safe for human consumption is called _____ water.

_____ 7. Domestic animals, such as livestock, do not require potable water.
A. True
B. False

_____ 8. The first step in the design of a treatment system is to conduct a test of the water.
A. True
B. False

_____ 9. A feed pump may be used to introduce _____ into the water
system.
   A. sodium hypochlorite
   B. ash
   C. magnesium
   D. fluoride

_____ 10. _____ involves the addition of large doses of chlorine to the water.

_____ 11. The _____ scale is used to quantify the level of acidity or alkalinity.

_____ 12. On the pH scale, a reading of 0 indicates the water is _____.
   A. extremely acidic
   B. neutral
   C. extremely alkaline
   D. moderately alkaline

13. Match the parts of the well with the proper part name.

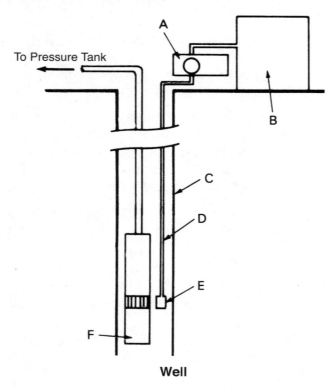

Well

Goodheart-Willcox Publisher

_____ a. Soda ash chlorine tank

_____ b. Well casing

_____ c. Chemical feeder tube

_____ d. Outlet near pump intake

_____ e. Pump motor

_____ f. Chemical feeder

Name _____

_____ 14. The treatment of water that is extremely acidic may require the
use of _____.
A. sodium hydroxide
B. soda ash
C. chlorine
D. None of the above.

_____ 15. The most trouble-free method of removing iron from water is with
a(n) _____ filter charged with potassium permanganate.
A. oxidizing
B. activated carbon
C. neutralized carbon
D. soda ash

_____ 16. Hydrogen sulfide is flammable and toxic.
A. True
B. False

_____ 17. _____ filters are cartridge filters that remove calcium and magnesium that produce deposits on piping, fixtures, and accessories.

_____ 18. Activated carbon (charcoal) filters are primarily used to remove
_____.
A. organic chemicals
B. chlorine
C. gases
D. All of the above.

_____ 19. A water _____ removes the dissolved calcium and magnesium by
ion exchange.
A. filter
B. softener
C. heater
D. strainer

_____ 20. The zeolite softening system process adds _____ to the water.
A. salt
B. magnesium
C. iron
D. None of the above.

_____ 21. The reverse osmosis filtering process is the fastest water treatment
process and wastes the least amount of water.
A. True
B. False

_____ 22. _____ is accomplished by heating the water above 160°F (71°C)
and retaining it at temperature for an extended period of time.

_____ 23. Bacteria can be effectively killed by ultraviolet light.
A. True
B. False

_____ 24. Because small particles in the water can shield bacteria from UV
light, a(n) _____ filter is necessary.
A. oxidizing
B. activated carbon
C. neutralized carbon
D. soda ash

_____ 25. Plumbers must rely on directions provided by the manufacturer,
the health department, and the local plumbing code.
A. True
B. False

Plumbing Fixtures

13

Objective: You will be able to recognize the materials from which fixtures are manufactured, identify various fixture types, and distinguish between the different types of toilets (water closets).

Instructions: *Carefully read Chapter 13 in the text and answer the following questions.*

1. List four common plumbing fixtures.

   _____

   _____

_____ 2. All plumbing fixtures are made of a single material, because combinations of materials are prohibited.
   A. True
   B. False

_____ 3. Porcelain produces a sanitary, easily cleaned surface.
   A. True
   B. False

_____ 4. Porcelain enamel is sometimes referred to as _____.
   A. "glass lining"
   B. kaolin
   C. vitreous enamel
   D. Both A and C.

_____ 5. _____ are used for washing hands and faces.
   A. Lavatories
   B. Bidets
   C. Sinks
   D. None of the above.

_____ 6. _____ are plumbing fixtures used for food preparation and dishwashing.

7. Match the lavatories and sinks with their proper names.

_____ a. One-piece molded type

_____ b. Ledge type

_____ c. Self-rimming type

_____ d. Built-in with metal rim

_____ e. Pedestal type

_____ f. Corner type

A

B

C

D

E

F

Elkay Mfg. Co.; Goodheart-Willcox Publisher; AnneIA/Shutterstock.com; Kohler Co.; American Standard Inc.

Name _____

_____   8. Touch-activated faucets are typically installed in _____ sinks.

_____   9. The drainpipe from a lavatory must be at least _____" in diameter.
   A. 1
   B. 1-1/4
   C. 1-1/2
   D. 2

_____   10. Sinks, laundry tubs, and garbage disposals must be connected to _____" waste piping.
   A. 1-1/4
   B. 1-1/2
   C. 2
   D. 2-1/2

_____   11. Plumbing fixtures must meet standards specifically established for each type of fixture.
   A. True
   B. False

12. List the abbreviations for three organizations that develop and maintain standards for plumbing fixtures.

_____

_____   13. Plumbers may be expected to install special grab bars and _____ in tubs and similar fixtures for individuals with physical disabilities.

_____   14. Scald-proof faucets are installed to prevent the user from being burned.
   A. True
   B. False

_____   15. Scald-proof faucets control the temperature of the water _____.

16. List three common sizes of shower bases.

_____

_____   17. Bathtubs that pump water and air through jets on the inside of the tub are called _____ bathtubs.

_____   18. A freestanding shower stall is generally installed in one piece.
   A. True
   B. False

_____   19. Shower stall floors are sloped _____" per foot to a center drain.

_____   20. Toilets are manufactured from _____ or reinforced plastic.
   A. cast-iron
   B. vitreous china
   C. polyglycol
   D. ABS-ACR

_____ 21. A(n) _____ is designed to carry away solid organic waste using water under pressure or gravity.

22. Match each toilet to its proper name.

_____ a. Pressure-assisted toilet

_____ b. Gravity-fed, rim-jet toilet

_____ c. Gravity-fed, siphon-jet toilet

A

B

C

Kohler Co.

_____ 23. Floor-mounted toilets are designed primarily for commercial buildings.
A. True
B. False

_____ 24. The older traditional toilets use about _____ gallons per flush.

_____ 25. The maximum amount of water a toilet can now consume is _____ gallon(s) per flush.
A. 1.1
B. 1.6
C. 2
D. 2.2

Name _____

_____ 26. The _____ is a companion fixture to the toilet, and is used for cleaning the perineal area of the body.

_____ 27. Urinals are installed in most residential bathrooms.
A. True
B. False

_____ 28. The trap of a(n) _____ urinal permits urine to enter a small reservoir through a layer of specially formulated oil that floats on top.

_____ 29. _____ sinks are found in custodial areas and other locations where cleaning equipment is serviced.

_____ 30. Water fountains used in public buildings are required to be wheelchair accessible.
A. True
B. False

Piping Materials and Fittings

Objective: You will be able to identify various pipes and fittings and describe their applications, grades, and sizes.

Instructions: *Carefully read Chapter 14 in the text and answer the following questions.*

_____ 1. Each plumbing system has _____ set(s) of piping.
A. one
B. two
C. four
D. six

_____ 2. The water supply system carries fresh water under pressure for _____.
A. drinking
B. cooking
C. bathing
D. All of the above.

_____ 3. The _____ system is one set of piping that carries away waste-water and solid waste.
A. PVC
B. DWV
C. ASTM
D. vent

_____
_____ 4. Pipe and fittings generally can be classified as either _____ or _____.

_____ 5. Nonpressure pipe is used for drainage, waste, and venting systems.
A. True
B. False

_____ 6. Plastic pipe and fittings dominate DWV systems in new residential and light commercial installations.
A. True
B. False

7. Identify the abbreviated types of plastic.

    a. ABS _____

    b. PVC _____

    c. CPVC _____

    d. PE _____

    e. PEX _____

    f. SR _____

8. Match each pipe type with its primary application.

Pipe	Primary Application
_____ a. ABS	A. Hot and cold water supply
_____ b. PEX	B. Hot water and some chemicals
_____ c. CPVC	C. Drain, waste, and vent
_____ d. PVC	D. Primarily pressure pipe
_____ e. SR	E. Septic tank, drain field, and storm drain
_____ f. PE	F. Natural gas

9. List three reasons why plastic pipe is so popular.

    _____

    _____

    _____

_____ 10. PVC pipe is suitable for hot water piping.
    A. True
    B. False

_____ 11. PEX is more likely to be damaged by freezing than other commonly used plastic materials.
    A. True
    B. False

_____ 12. _____ is manufactured for distributing cold and/or hot water inside a structure.
    A. Schedule 40 PVC
    B. Schedule 80 PVC
    C. PEX
    D. All of the above.

Name _____

13. Identify the abbreviations listed below.

a. OD _____

b. ID _____

c. WT _____

d. SDR _____

_____ 14. Closet flanges connect the water closet to the drain system.
           A. True
           B. False

_____ 15. Steel pipe is used for _____.
           A. hot and cold water
           B. gas and air piping
           C. steam and hot water
           D. All of the above.

_____ 16. The standard length of steel pipe is _____'.

_____ 17. Pressure fittings are suitable for drainage systems.
           A. True
           B. False

_____ 18. The _____ is a fitting that is *not* used to change direction.
           A. elbow
           B. coupling
           C. tee
           D. drop elbow

19. List the three grades of steel pipe.

_____

_____

_____

_____ 20. The threads on iron pipe fittings are _____ to form a watertight connection.

21. Match each type of copper pipe with its color code.

	Pipe	Color Code
_____	a. K copper	A. Blue
_____	b. L copper	B. Yellow
_____	c. M copper	C. Green
_____	d. DWV copper	D. Red

_____ 22. The weight of copper water pipe refers to the _____ of the wall.

_____ 23. The outside diameter of copper pipe is _____" larger than the standard designation.
   A. 1/32
   B. 1/16
   C. 1/8
   D. 3/16

24. List the two types of solder fittings for copper pipe.

_____

_____ 25. Wrought copper solder fittings are available in a greater variety of shapes than cast fittings.
   A. True
   B. False

_____ 26. Copper drainage fittings are available with inlets ranging from 1-1/2" to 4" in diameter.
   A. True
   B. False

_____ 27. Quick-connect fittings _____.
   A. are available in both plastic and brass
   B. require special tools to install
   C. can be used to join PEX pipe up to 2" in diameter
   D. can be used to install only cold water supply piping

28. Match the fittings with the proper name.

A   B   C   D

NIBCO, Inc.

_____ a. Union

_____ b. 45° ell

_____ c. Adapter

_____ d. 90° ell

_____ 29. (A)n _____ is a piece of pipe, 12" or less in length, that is threaded on both ends.

30. What has been the primary use of vitrified clay pipe?

_____

_____ 31. _____ fittings make a watertight seal when a brass sleeve is squeezed between the nut, pipe, and the body of the fitting.

Valves and Meters

Objective: You will be able to identify and list the types, applications, and construction of various valves, faucets, and meters.

Instructions: *Carefully read Chapter 15 in the text and answer the following questions.*

1. List the six materials that valves are generally made of.

_____

_____

_____ 2. Valves are available in standard sizes ranging from _____" to _____" diameter.
   A. 1/4; 8
   B. 1/4; 12
   C. 3/4; 8
   D. 3/4; 12

_____ 3. Valves may *not* be necessary in a DWV piping system.
   A. True
   B. False

_____ 4. The stop valve has remained unchanged for nearly 100 years.
   A. True
   B. False

_____ 5. Three-way ground key valves are used in _____.
   A. water pumps
   B. gas burner units
   C. automobile washing equipment
   D. All of the above.

_____ 6. Corporation valves are available in 1/2" through 2" diameters.
   A. True
   B. False

_____ 7. A meter stop allows the water to be shut off at the inlet to the
_____.

    A. valve
    B. corporation valve
    C. meter
    D. building

8. Match valve parts with the proper part name.

    _____ a. Inlet

    _____ b. Outlet

    _____ c. Valve seat

    _____ d. Washer

    _____ e. Handle

    _____ f. Packing nut

    _____ g. Packing box

    _____ h. Bonnet

    _____ i. Stem

    _____ j. Screw thread

Valve

William Powell Co.

_____ 9. A _____ valve allows water on the downstream side to be drained.
    A. stop and waste
    B. gate
    C. globe
    D. ball

_____ 10. A(n) _____ valve gets its name from the gate-like disk that slides
across the path of the flow.

_____ 11. A(n) _____ valve is open with just one-quarter turn and provides
little resistance to flow.

_____ 12. Plumbing fixtures that use flush valves do *not* need a storage tank.
    A. True
    B. False

_____ 13. There is no advantage to having flush valves or gravity flow
storage tanks.
    A. True
    B. False

Name _____

_____ 14. Flush valves are available in the _____ type and _____ type.
  A. piston; gravity
  B. diaphragm; piston
  C. gate; trip lever
  D. piston; ballcock

_____ 15. A control stop is generally installed in the water supply line serving the flush valve.
  A. True
  B. False

_____ 16. A _____ is installed between the outlet of the flush valve and fixture to prevent back siphoning.
  A. circuit breaker
  B. diaphragm
  C. vacuum breaker
  D. plunger

_____ 17. A check valve allows flow in only one direction and is used to prevent backflow.
  A. True
  B. False

_____ 18. A _____ valve allows excess pressure to bleed off to the atmosphere.
  A. float
  B. pressure-relief
  C. pressure regulator
  D. blow down

19. Identify the parts of the float-controlled valve.

  a. _____

  b. _____

  c. _____

Fluidmaster, Inc.

_____ 20. A(n) _____ reduces water pressure in a building.
    A. stop valve
    B. angle valve
    C. pressure regulator
    D. relief valve

_____ 21. A pressure regulator valve is installed if the water pressure exceeds _____.
    A. 40 psi
    B. 60 psi
    C. 80 psi
    D. 100 psi

_____ 22. Backwater valves are a type of check valve designed for DWV piping systems.
    A. True
    B. False

_____ 23. A(n) _____ is a mechanical device that measures the amount of water passing through the water service pipe into a building.

_____ 24. The water meter is usually the property of the _____.
    A. landowner
    B. city
    C. plumbing contractor
    D. None of the above.

_____ 25. Water meters are manufactured with capacities of _____ gallons per minute.
    A. 15, 20, or 25
    B. 20, 25, or 30
    C. 20, 30, or 50
    D. 25, 35, or 45

Name _____ Date_____

Period_____ Course _____

Score_____

Water Heaters

16

Objective: You will be able to identify the two types of heating/storage tanks, describe their differences, demonstrate how their controls work, and explain the steps for water heater installation.

Instructions: *Carefully read Chapter 16 in the text and answer the following questions.*

_____ 1. The _____ is responsible for installing the piping for water heaters.
   A. plumber
   B. electrician
   C. pipe fitter
   D. HVAC contractor

_____ 2. Storage-type water heaters are the most common water heating equipment.
   A. True
   B. False

3. Match the operational parts and processes of the water heater with the proper name.

   _____ a. Thermostat

   _____ b. Temperature/pressure relief valve

   _____ c. Flue

   _____ d. Cold water inlet

   _____ e. Insulation

   _____ f. Hot water rises

   _____ g. Heating unit

   _____ h. Cold water sinks to the bottom of tank

   _____ i. Overflow piping

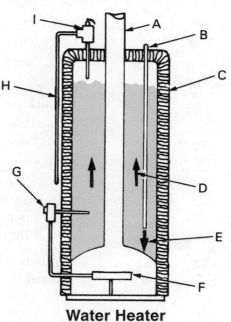

Water Heater

Goodheart-Willcox Publisher

_____   4. In a storage-type water heater, the cold water enters near the
                         _____ of the tank.
                         A. center
                         B. top
                         C. bottom

_____   5. In a storage-type water heater, hot water is drawn off near the
                         _____ of the tank.
                         A. top
                         B. bottom
                         C. center

_____   6. When water is heated, it _____.
                         A. contracts
                         B. expands
                         C. condenses

7. What are the two energy sources that are most used for water heaters?

_____

_____

_____   8. Liquefied petroleum is the most common energy source for water
                         heaters.
                         A. True
                         B. False

_____   9. Electric water heaters use an insulated heating element to heat the
                         water.
                         A. True
                         B. False

_____   10. Most electric water heaters require _____V wiring.
                         A. 240
                         B. 120
                         C. 220
                         D. None of the above.

_____   11. In normal operation, only the upper electric heating element is
                         working.
                         A. True
                         B. False

_____   12. The _____ is the speed at which cold water can be heated.
                         A. recovery rate
                         B. heated rate
                         C. energy efficiency
                         D. None of the above.

_____   13. When selecting a storage-type water heater, the _____ should be
                         considered.
                         A. capacity of the water heater
                         B. fuels available
                         C. ability of the tank to hold heat
                         D. All of the above.

Name _____

_____ 14. The Energy Guide Program is administered by the _____.
  A. Federal Trade Commission
  B. American Gas Association
  C. American Water Worker
  D. Underwriters Lab

_____ 15. A(n) _____ turns the energy source of the water heater on and off as needed.

16. Match the parts of the valve control unit with the proper part name.

_____ a. Gas supply line

_____ b. Solenoid valve

_____ c. Solenoid coil

_____ d. Heat-sensing element

_____ e. Pilot

_____ f. Pilot supply line

_____ g. Thermocouple

_____ h. Burner supply line

_____ i. Thermostat valve

_____ j. Flame

Valve Control Unit

Goodheart-Willcox Publisher

_____ 17. The _____ thermostat cuts off the current to the solenoid should the water temperature exceed a predetermined safe limit.

_____ 18. The high-limit protector is an automatic safety device that shuts off all electrical current if the water temperature exceeds _____.
  A. 180°F
  B. 180°C
  C. 82°C
  D. Both A and C.

_____ 19. The temperature/pressure (T/P) relief valve is installed in a specially designed opening in the top of the tank, and prevents excess temperature and pressure buildup in the tank.
  A. True
  B. False

_____ 20. The major hazard of water heaters is related to the buildup of _____ water in the tank.
  A. superheated
  B. warm
  C. cold
  D. potable

_____ 21. When water changes to steam, its volume increases approximately _____ times.

_____ 22. The tank can rupture if a _____ has not been installed.
A. gate valve
B. relief valve
C. mixing valve
D. tempering valve

23. What should be done if water does not flow out of the drip line when the T/P relief valve is manually opened to check it?

_____

_____

_____ 24. Gas-fired water heaters should *not* be placed near a chimney or flue.
A. True
B. False

_____ 25. Directly connecting a gas-fired water heater to a furnace flue is often permitted if a T connector is used.
A. True
B. False

_____ 26. For a higher volume of hot water, two or more water heaters may be connected in _____.

_____ 27. When using a solar collector to heat water, it is typical to install _____.
A. two water heaters in parallel
B. a smaller water heater
C. two water heaters in series
D. None of the above.

28. What type of water heater is most frequently used in the United States for point of use heating of water?

_____

_____ 29. In order to correctly size an instantaneous hot water heater, accurate estimates of _____ must be made.

_____ 30. An important consideration with instantaneous water heaters is _____.
A. energy source
B. installation cost
C. low flow rate
D. Both B and C.

Name _____

31. Match each of these common fixtures with its hot water demand.

		Fixture	Hot Water Demand
_____	a.	Kitchen sink	A. 3.6
_____	b.	Dishwasher	B. 1.5 GPM
_____	c.	Bathtub	C. 1.6 GPM
_____	d.	Shower	D. 0.3 GPM
_____	e.	Lavatory	E. 2.5 GPM
_____	f.	Washing machine	F. 3.3 GPM

_____ 32. Gas- or oil-fired instantaneous units require a vent to the exterior of the building.
A. True
B. False

_____ 33. Heating water with solar collectors does little to reduce the energy needed to heat water.
A. True
B. False

_____ 34. Heat pumps extract heat from the _____ air and _____ it to the
_____ water.

_____ 35. Hot water pipes from the water heater to each fixture should be _____ to reduce energy consumption.

Designing Plumbing Systems

Objective: You will be able to determine the size of drainage and water supply piping, design efficient and serviceable plumbing systems, and plan plumbing installations to serve disabled individuals. You will be able to recognize and avoid cross connections and describe storm water collection and drainage systems.

Instructions: *Carefully read Chapter 17 in the text and answer the following questions.*

_____ 1. When designing a plumbing system, only the architect can identify basic needs for the building.
A. True
B. False

_____ 2. In commercial buildings, the number of people who use the restroom facilities must be considered.
A. True
B. False

_____ 3. To reduce costs, rooms needing plumbing should be placed near each other.
A. True
B. False

_____ 4. It is common practice to arrange rooms needing plumbing _____.
A. back-to-back
B. above and below each other
C. at all corners of the building
D. Both A and B.

_____ 5. It is *not* good practice to have one soil stack serving more than one room.
A. True
B. False

_____ 6. By stacking bathrooms and kitchens in a multistory building, all plumbing lines can be placed in a single _____.

_____ 7. The abbreviation ADA stands for _____.

_____ 8. The ADA requires at least one handicapped accessible restroom on each floor.
   A. True
   B. False

9. While planning a kitchen, a planner must be concerned with the work area "triangle" created by what three items?

_____

_____

_____ 10. Sinks, dishwashers, and garbage disposals may need a water supply and/or a drainage pipe.
   A. True
   B. False

_____ 11. The _____ dimensions indicate where the DWV and water supply must be located in the wall and/or floor.

_____ 12. The _____ pipe system operates on gravity.

_____ 13. Because DWV piping systems operate on gravity, a drop of _____"
to _____" per foot is considered adequate.
   A. 1/8; 1/2
   B. 1/4; 1/2
   C. 1/2; 3/4
   D. 3/4; 2

_____ 14. A DWV pipe that is too large is _____.
   A. likely to clog
   B. extremely expensive
   C. more difficult to install
   D. All of the above.

_____ 15. There are approximately _____ gallons of water in one cubic foot.
   A. 7.5
   B. 9
   C. 3.25
   D. 8.5

_____ 16. A lavatory that discharges _____ gallons of water per minute has a load factor of 1.

_____ 17. Stacks must never be smaller than the largest branch entering them.
   A. True
   B. False

_____ 18. _____ permit air to circulate through the waste piping system.

Name _____

_____ 19. A _____ in the waste line prevents sewer gas from entering the building by providing a water seal.
A. trap
B. cleanout
C. vent
D. None of the above.

20. Match the ways a trap can lose its water seal with its description.

Type of Water Seal Loss	Description
_____ a. Siphonage	A. Downdraft in stack
_____ b. Back pressure	B. Foreign material caught in trap
_____ c. Evaporation	C. System not in use for extended period
_____ d. Capillary action	D. Water drawn from one fixture by discharge of another
_____ e. Wind	E. Air pressure builds in system

_____ 21. _____ venting is the best venting method because it vents every trap separately.

_____ 22. The fittings used to install vent piping are the same as those used for the waste piping.
A. True
B. False

_____ 23. The water supply system is designed to operate at _____ psi within the building.

_____ 24. _____ pipe is not suitable for installation of the water supply system.
A. Black iron
B. CPVC
C. Lead
D. Both A and C.

_____ 25. In a water tower, the pressure increases _____ psi for each foot of height.

26. Using the tower diagram, find the water pressure for each floor.

_____ a. Basement

_____ b. 1st floor

_____ c. 2nd floor

_____ d. 3rd floor

_____ e. 4th floor

_____ f. 5th floor

_____ g. 6th floor

Tower Diagram

_____ 27. Consideration must be given for pressure loss through fittings and long lengths of runs.
A. True
B. False

_____ 28. Water piping must be protected from freezing temperatures because the pipe may _____.

_____ 29. The number of fittings installed has no effect on water pressure.
A. True
B. False

_____ 30. Air gap applications are used to prevent cross connections.
A. True
B. False

_____ 31. A(n) _____ valve is used to prevent a water heater from exploding if the thermostat ceases to function.

_____ 32. Shutoff valves allow the plumber to _____ small parts of the piping system.
A. clean
B. isolate
C. supply
D. None of the above.

_____ 33. Piping systems containing potable water can be interconnected with other piping systems.
A. True
B. False

Name _____

_____ 34. _____ is caused by the sudden stop of water flow.

_____ 35. _____ are installed in water piping to prevent water hammer.
    A. Relief valves
    B. Air chambers
    C. Bypass valves
    D. Pressure reducing valves

Preparing for Plumbing System Installation

18

Objective: You will be able to describe steps involved in locating and bringing in DWV and water supply in a building. You will be able to use the prescribed techniques for working with plumbing materials and fixtures and describe methods of laying out plumbing systems.

Instructions: *Carefully read Chapter 18 and answer the following questions.*

1. Using the numbers 1, 2, 3, and 4, place stages for installing plumbing in proper order.

    _____ a. Sink, lavatory, and showerhead installation

    _____ b. Piping enclosed in the walls

    _____ c. Sewer and water supply installation

    _____ d. Final inspection

_____ 2. _____ are required to make connections to the water main and the sanitary sewer.
  A. Mechanical tees
  B. Adapters
  C. Permits
  D. None of the above.

_____ 3. If permitted by local code, water supply and building sewer pipes can be run in a(n) _____ trench that provides a minimum distance between the pipes.

_____ 4. Once the building sewer is installed, it must be inspected before backfilling.
  A. True
  B. False

_____ 5. The two different methods used to attach the corporation stop to the water main are to drill and tap the side wall of the main or to attach a(n) _____ to the main.

_____ 6. The _____ has responsibility for installing the water line from the curb stop into the house.
   A. plumbing contractor
   B. city workers
   C. architect
   D. None of the above.

_____ 7. A(n) _____ is a valve installed near the point where the water supply line enters the building so a water meter can be attached.
   A. angle meter valve
   B. curb stop
   C. curb box
   D. None of the above.

_____ 8. The _____ is the drainage piping between the building's foundation and the sewer main or septic tank.

_____ 9. Allowances for finished walls and flooring are an important part of the second rough.
   A. True
   B. False

_____ 10. A curb box is a control valve installed between the corporation stop and the structure.
   A. True
   B. False

_____ 11. With cast-in-place concrete floors above ground, _____ should be used to produce opening for pipes to pass from floor-to-floor.

_____ 12. _____ strips may be needed to make a wall wide enough to conceal piping.

_____ 13. A(n) _____ is installed in the wall to support the weight of a wall-hung lavatory.
   A. 2 x 6 block
   B. cast-iron pipe
   C. extra stud
   D. large nails

_____ 14. When possible, the stack should be installed directly behind the _____.
   A. lavatory
   B. tub
   C. toilet
   D. shower stall

_____ 15. When cutting floor joists, _____ joists and headers ensures that the strength of the floor remains.

Name _____

_____ 16. The water supply piping is installed after the _____ system is completed.
A. gas
B. PVC
C. ABS
D. DWV

_____ 17. The cold water supply for the toilet will always come up through the floor.
A. True
B. False

_____ 18. The lavatory water supply piping, like the toilet, should stub-out through the wall.
A. True
B. False

_____ 19. Commercial buildings often have wall-hung _____.
A. toilets
B. urinals
C. lavatories
D. All the above.

_____ 20. After all layout _____ have been rechecked for accuracy, the drilling and cutting of openings can begin.

Name _____ Date _____

Period_____ Course _____

Score_____

DWV Pipe and Fitting Installation

19

Objective: You will be able to describe procedures for locating and installing DWV piping in a building. You will be able to use the prescribed techniques for working with plumbing materials, use three methods of measuring pipe, and describe methods of testing and inspecting plumbing systems.

Instructions: *Carefully read Chapter 19 in the text and answer the following questions.*

_____ 1. The building sewer and water supply are installed during the _____.
        A. first-rough
        B. second-rough
        C. third-rough
        D. final inspection

2. List the three types of measurements taken to determine length of pipe.

_____

_____

_____

_____ 3. The _____ method of determining length of pipe is useful only with DWV fittings.

_____ 4. The plumber must know the fitting _____ for the type and size of fitting being used.

_____ 5. The _____ is the amount of a run of a pipe that is taken up by the fitting.

_____ 6. _____ codes include many detailed requirements for DWV piping.

_____ 7. A _____ will allow the dimensions to be checked before final assembly.
        A. tee
        B. trial assembly
        C. standpipe
        D. None of the above.

_____   8. Most plumbing codes require _____ at each change in direction of the DWV piping greater than 45°.

_____   9. The amount of the stack that extends above the roof is specified by code.
       A. True
       B. False

_____   10. In cold climates, larger pipes are used above the insulated area of the building to prevent the vent from _____.

_____   11. _____ is installed on the vent pipe at the roof line to prevent roof leaks around the pipe.

12. What is the main concern when installing horizontal runs of drain/waste piping?

_____

_____

13. List the four ways bathtubs and showers are installed.

_____

_____

_____   14. DWV plastic piping is generally measured using the direct and indirect method.
       A. True
       B. False

_____   15. A(n) _____ should be used when cutting PVC pipe to ensure that the cut is square.

_____   16. Horizontal runs of plastic pipe should be supported every _____' to _____'.
       A. 2; 4
       B. 3; 4
       C. 4; 6
       D. 6; 8

_____   17. Copper DWV pipe measurements are generally taken center-to-center.
       A. True
       B. False

_____   18. When cutting copper pipe with a pipe cutter, the cut end of the pipe should be reamed.
       A. True
       B. False

Name _____

_____ 19. Copper DWV pipe and fittings are joined by _____.

_____ 20. Horizontal runs of copper pipe should be supported every _____′ to _____′.
A. 2; 4
B. 3; 4
C. 4; 6
D. 6; 8

_____ 21. Black iron pipe is used for DWV piping.
A. True
B. False

_____ 22. Black iron pipe should not be used for _____ and compressed air systems.

_____ 23. When threading pipe, _____ should be used to lubricate the die.

_____ 24. _____ should be applied to the pipe threads before joining the pipe and fitting.

_____ 25. Galvanized and black steel pipe should be supported every _____′ to _____′.
A. 2; 4
B. 3; 4
C. 4; 6
D. 6; 8

_____ 26. Cast-iron pipe is generally cut with a compound lever or hydraulic pipe cutter.
A. True
B. False

_____ 27. No-hub cast-iron pipe is joined with a(n) _____ and a stainless steel clamp.

_____ 28. Bell and spigot cast-iron pipe is joined with neoprene compression gaskets or _____.

_____ 29. _____ is packed into the bell to form a seal before pouring the lead.

_____ 30. When pouring a cast-iron joint with a 4″ diameter, about _____ pounds of lead is melted.
A. 2
B. 4
C. 8
D. 12

_____ 31. When molten lead strikes _____ surfaces, steam is created that will expand and harm the joint.

_____ 32. A(n) _____ creates a dam around the top of the bell when pouring lead into horizontal runs.

_____ 33. Lead and oakum joints are seldom used today in residential or light commercial plumbing.
   A. True
   B. False

Installing Water Supply Piping

20

Objective: You will be able to explain the methods of measuring pipe between fittings, describe proper procedures for locating and installing water supply systems, use prescribed techniques for working with pipes and fittings made from PVC, CPVC, PEX, and copper.

Instructions: *Carefully read Chapter 20 in the text and answer the following questions.*

_____ 1. The second rough stage of plumbing installation includes _____ covered by finished wall material.
   A. DWV piping
   B. water piping
   C. Both A and B.
   D. None of the above.

_____ 2. The water supply system can be installed after the _____ piping system has been completed.

3. List at least six questions that might be asked before routing the water supply piping.

_____

_____

_____

_____

_____

_____

_____

_____ 4. Water pressure at the meter can be determined by the local water authority.
   A. True
   B. False

_____ 5. Proper sizing of the water supply piping is *not* critical to the oper-
ation of the plumbing system.
A. True
B. False

6. List three interrelated factors that the size of the pipe and fittings are dependent upon.

_____

_____

_____

_____

_____ 7. Local code regulates pipe size.
A. True
B. False

_____ 8. The water filter is located on the _____ side of the meter.
A. street
B. curb
C. building
D. None of the above.

_____ 9. Fluctuation of the cold water pressure at the showerhead could
result in a significant change in water _____.

_____ 10. When sizing the water supply pipe for the kitchen and the
laundry facilities, the _____ should be considered first.

11. Using the model code provided with the text, list the flow rates for the following fixtures:

a. Lavatory faucet _____

b. Showerhead _____

c. Sink faucet_____

d. Toilet _____

_____ 12. To ensure that work passes inspection, plumbers must be thor-
oughly familiar with the _____.
A. model code
B. international code
C. standard code
D. local code

_____ 13. Water hammer can cause pipes to vibrate and possibly burst.
A. True
B. False

Name _____

_____ 14. Many plumbers install _____ at each stub-out to act as water hammer arrestors.

_____ 15. Air chambers may not meet the code requirement for water hammer arrestors near quick closing valves.
A. True
B. False

_____ 16. Valves for individual fixtures are installed during the _____ stage.
A. first-rough
B. second-rough
C. third-rough
D. finish

_____ 17. Code requires that a valve be installed on the _____ pipe entering the water heater.

_____ 18. The cold water inlet of a water heater is fitted with a(n) _____ that directs entering cold water to the bottom of the tank.

_____ 19. PVC should be used for the connections to the water heater.
A. True
B. False

_____ 20. Joints between copper and galvanized pipe should be made with a(n) _____ fitting.

_____ 21. The cold water stub-out is always installed on the _____ side while the hot stub-out is always on the _____ side.

_____ 22. Plumbing for a clothes washer is installed inside a finished wall, with the pipes connected to a washing _____ box.
A. inlet
B. outlet
C. square
D. None of the above.

_____ 23. PVC is used for cold water, while _____ is used for hot water.

_____ 24. PVC is _____ in color and CPVC is _____ in color.

_____

_____ 25. Expansion loops should be installed near the _____ of long runs of pipe.

_____ 26. When joining PVC pipe and fittings with solvent cement, allow _____ hours before pressure testing.

_____ 27. When copper pipe and tubing is installed, the _____ measurements are frequently used.
  A. face-to-face
  B. face-to-center
  C. back-to-back
  D. center-to-center

_____ 28. Copper pipe and tubing should be cut with a(n) _____.

_____ 29. The cut end of a pipe is reamed to remove the _____.

_____ 30. Installing PEX differs in several ways from installing rigid pipe such as PVC, CPVC, and copper.
  A. True
  B. False

_____ 31. PEX joints are made with _____ rings and a special _____ tool.
_____

_____ 32. PEX pipe expands and contracts _____" for every _____' of tube for every 10°F change in temperature.
  A. 1; 100
  B. 2; 200
  C. 3; 300
  D. All of the above.

Supporting and Testing Pipe

21

Objective: You will be able to install appropriate supports for residential and light commercial piping systems, install anchors in concrete and masonry, and test both DWV and water supply piping systems.

Instructions: *Carefully read Chapter 21 in the text and answer the following questions.*

_____ 1. Both DWV and water supply piping must be supported both
_____   _____ and _____.

2. List the three functions of horizontal supports.

   _____

   _____

   _____

3. List the two functions of vertical supports.

   _____

   _____

_____ 4. Some of the needed supports are installed as pipe and fittings are
                                put in place.
                                A. True
                                B. False

_____ 5. Plumbing code requires that piping be supported to maintain
_____   alignment, prevent sagging, and to allow for _____ and _____.

6. List three important factors in making the selection of pipe hangers and supports.

   _____

   _____

   _____

7. List three ways in which copper pipe and fittings can be supported horizontally.

_____

_____

_____

_____ 8. Pipe straps are generally used for horizontal support.
   A. True
   B. False

_____ 9. When supporting plastic pipe, hangers that permit expansion and contraction movement should be used.
   A. True
   B. False

_____ 10. Long horizontal runs of DWV piping must be supported in a way that maintains a constant _____.

_____ 11. Supporting the weight of DWV vertical pipe is accomplished by securing _____ at each floor level.

_____ 12. Screws and glue are used to secure pipe hangers and supports to wood-framed structures.
   A. True
   B. False

_____ 13. A(n) _____ is often used to make pipe supports for parallel pipe runs.

_____ 14. In steel-framed buildings, clamp-like devices are often used to attach hangers for smaller pipes.
   A. True
   B. False

15. Explain why it is advantageous to use stud-type anchors.

_____

_____

_____

_____

_____ 16. Concrete anchors may use _____ to secure objects.
   A. hex head bolts
   B. machine screws
   C. hex nuts
   D. All of the above.

_____ 17. The holes in an object being mounted can be used to guide the masonry bit.
   A. True
   B. False

Name _____

18. Match each anchor with its proper name.

_____ a. Stud

_____ b. Drop-in anchor

_____ c. Self-drilling

_____ d. Insert

_____ e. Full thread sleeve

Anchors

ITW Ramset/Red Head; Phillips Drill Co.

19. Define the following terms.

Lag shield: _____

_____

Caulking anchors: _____

_____

Toggle bolts: _____

_____

20. Match the pipe hangers and supports with their proper names.

_____ a. Pipe clamp

_____ b. I-beam clamp

_____ c. Beam clamp

_____ d. Steel "C" clamp

_____ e. Pipe strap

_____ f. Swivel loop hanger

_____ g. Wall bracket

_____ h. Clevis hanger

Pipe Hangers and Supports

Goodheart-Willcox Publisher; Modern Hanger Corp.; NIBCO, Inc.

_____ 21. _____ anchors provide a fastener that is permanently attached to the concrete or masonry and is internally threaded to accept machine screws or bolts.

_____ 22. All equipment necessary to conduct tests for a rough-in inspection must be supplied by the plumber.
A. True
B. False

_____ 23. A minimum of _____' of water head must be used to water test DWV piping.
A. 15
B. 10
C. 5
D. 3

_____ 24. A(n) _____ may be used to produce pressures as high as 400 psi.

_____ 25. When air testing a DWV pipe system, a pressure of _____ psi is generally adequate.
A. 20
B. 15
C. 10
D. 5

Installing Fixtures, Faucets, and Appliances

22

Objective: You will be able to describe proper installation procedures for each fixture, faucet, and appliance presented in this chapter. Identify the specials tools needed to install the various fixtures and procedures for connection to the water supply and DWV connections.

Instructions: *Carefully read Chapter 22 in the text and answer the following questions.*

_____ 1. Due to the unique characteristics of the various fixtures, plumbers should study the manufacturer's instructions before attempting to install any fixture.
A. True
B. False

_____ 2. Mounting the _____ before installing the lavatory or sink makes it much easier to reach and tighten the nuts on the underside.
A. rim
B. faucet
C. P-trap
D. cleanout

_____ 3. Water supply tubes are made from flexible PVC tubing, braided stainless steel, and _____ soft copper tubing.

_____ 4. The adapter fittings for the water supply and drain piping stub-outs should be installed before putting the lavatory or sink in place.
A. True
B. False

_____ 5. Lavatories are typically _____" above the floor.

_____ 6. When self-rimming and metal-rimmed lavatories and sinks are installed, adhesive caulk is used to seal the joint at the countertop.
A. True
B. False

_____ 7. According to the code, the minimum diameter for the fixture tail-piece of a lavatory is _____.
A. 1"
B. 1-1/4"
C. 1-1/2"
D. 2"

_____ 8. The _____ body is installed after the sink or lavatory is placed in position.

_____ 9. _____ is used to seal the joint between the strainer body and the sink or lavatory.

_____ 10. A strainer wrench is used to prevent the strainer body from rotating, while a(n) _____ wrench is used to tighten the nut.

_____ 11. Kitchen sinks are frequently fitted with a(n) _____.
A. garbage disposal
B. ballcock
C. indirect waste
D. None of the above.

_____ 12. The outlet of a garbage disposal is connected directly to the DWV piping with _____ and fittings.

_____
_____ 13. The cold water supply tube is connected to the _____-side angle valve and the hot water supply tube is connected to the _____-side angle valve.

_____ 14. If a double- or triple-bowl sink is being installed, a(n) _____ fitting will be needed at the P-trap inlet.
A. elbow
B. tee
C. sleeve
D. check valve

_____ 15. Shower drains are connected to the DWV piping during the _____.
A. first-rough
B. second-rough
C. third-rough
D. last-rough

_____ 16. Installation of a shower valve is essentially the same as installation of a tub/shower valve.
A. True
B. False

_____ 17. The shower valve has an outlet for the tub spout.
A. True
B. False

_____
_____ 18. During the second-rough, the tub/shower valve is secured to the frame of the building and connected to the _____ and _____ water piping.

Name _____

_____ 19. After installing the showerhead and tub spout, all connections should be checked for _____.

_____ 20. _____ bathtubs pump water and air through jets on the inside of the tub.

_____
_____ 21. The two types of toilet installation are _____ and _____.

_____ 22. Most codes require a _____ at the toilet stub-out to allow shutting off the water in case of a malfunction.
A. flange
B. check valve
C. valve
D. cleanout

_____ 23. The connection between a toilet and the DWV piping system is made with a(n) _____ flange.

_____ 24. When installing a closet bowl, place the bowl temporarily over the closet flange to check for _____.

25. Number the following in order of the steps for installing a toilet tank.

_____ a. Place a spud washer over the water inlet of the bowl.

_____ b. Attach the flush lever to the flush tank.

_____ c. Install float rod and float on the flush valve if necessary.

_____ d. Carefully place the tank in position so the opening fits over spud washer.

_____ e. Secure the ballcock assembly to bottom of the tank.

_____ f. Secure with tank bolts.

_____ 26. Once the tank has been installed and secured, the plumber must check for leaks.
A. True
B. False

_____ 27. When correcting the water level in the tank, the _____ on the ball-cock is adjusted.

_____ 28. A(n) _____ is used for personal hygiene in cleaning the perineal area of the body.

_____ 29. Urinals are commonly installed in _____ restrooms.

_____ 30. A flush valve is always used in conjunction with a(n) _____ and
_____ _____.

_____ 31. _____ sinks are typically made of stainless steel or cast iron, and are usually located in a custodian's room.

_____ 32. In schools, hospitals, and office buildings, drinking water that is not cooled is supplied by _____.

_____ 33. Many refrigerators are equipped with icemakers that require a _____ water supply tubing.
A. 1/4"
B. 1/2"
C. 3/4"
D. None of the above.

_____ 34. A(n) _____ valve can be used to connect the water supply pipe and an icemaker or a humidifier.

_____ 35. Enclosed bath and shower stalls may be equipped with _____ that heat the water and dispense controlled amounts of steam into the stall.

Septic Systems

23

Objective: You will be able to explain the operation of a simple septic system, list the essential materials and describe methods used in construction of a septic tank and leach field, and describe the construction and operation of alternative systems.

Instructions: *Carefully read Chapter 23 in the text and answer the following questions.*

_____ 1. The Babylonians constructed the first known sewer system more than _____ years ago.

_____ 2. _____ officials generally have jurisdiction over the installation of private waste disposal systems.
   A. Local plumbing
   B. Local health
   C. State health
   D. City inspector

3. List the two essential parts of a private waste disposal system.

   _____

   _____

4. Number the following in order of the steps of the disposal process.

   _____ a. Solids settle to the bottom.

   _____ b. Liquid leaves the septic tank through sealed pipe and enters distribution box.

   _____ c. Waste enters the leach field through perforated pipe.

   _____ d. Waste enters the septic tank.

   _____ e. Wastewater leaves the distribution box.

   _____ f. Water is absorbed into the ground.

   _____ g. Sealed pipe carries liquid waste to the leach field.

_____ 5. Leach beds are commonly found in swampy areas or where flooding is common.
   A. True
   B. False

_____ 6. A modern home produces about _____ gallons of waste daily for
each person living in the house.

_____ 7. Plantings over leach beds should generally be limited to _____.

_____ 8. The size of the _____ and the amount of gravel placed in the leach
field runs affect the capacity of the system.

_____ 9. Water entering the ground from the leach field must *not* pollute
wells.
A. True
B. False

_____ 10. It is generally accepted that leach fields should be installed on
ground that is of _____ the house.
A. lower elevation than
B. higher elevation than
C. equal elevation to

_____ 11. The _____ tank is probably the most common septic tank.
A. steel
B. fiberglass reinforced
C. concrete
D. None of the above.

_____ 12. The purpose of a(n) _____ is to provide a place where the many
lines of the leach field connect.

_____ 13. It is impossible to have a completely closed waste disposal system
that recycles the flush water.
A. True
B. False

_____ 14. An aeration wastewater treatment plant injects _____ into the
sewage, causing the growth of aerobic bacteria.

_____ 15. The Clivus Multrum eliminates the need for flush water.
A. True
B. False

_____ 16. A closed system of waste treatment was developed primarily for
portable toilets.
A. True
B. False

17. What is *dosing*?

_____

_____

_____

_____

Name _____

_____ 18. Strong chemicals used for cleaning drain lines should *not* be used with a septic tank because these chemicals kill _____ that are essential for it to function.
A. bacteria
B. grey water
C. solids
D. ozone

19. List three categories of products that should *not* be introduced in a septic system.

_____

_____

_____ 20. Minimizing the amount of liquid that enters the septic system gives the bacteria additional time to break down the waste.
A. True
B. False

21. List five practices that can reduce the amount of liquid entering the septic system.

_____

_____

_____

_____

_____

_____ 22. When a high volume of waste enters the septic system in a short period of time, insufficient time is available for the waste to _____.

_____ 23. It is generally recommended that the tank be pumped at _____-year intervals.

_____ 24. If the septic tank has a filter at the outlet, the filter should be cleaned regularly.
A. True
B. False

Name _____ Date _____

Period_____ Course _____

Score_____

Storm Water and Sumps

Objective: You will be able to explain the difference between storm sewers and sanitary sewers, install drainage piping that collects storm water runoff, and describe the function and installation of a sump pump.

Instructions: *Carefully read Chapter 24 in the text and answer the following questions.*

1. List the two separate drain/waste piping systems installed in a typical building.

   _____

   _____

_____ 2. Some older buildings may have a combined DWV and storm
water piping system.
A. True
B. False

_____ 3. Only water that occurs naturally on the property is collected by
the _____ system.

_____ 4. The _____ system collects water from streets, parking lots, and
each building in the service area.

_____ 5. It is generally preferable to pipe storm water to a sump rather
than to the surface.
A. True
B. False

_____ 6. The _____ collects groundwater and channels it into a basin called
a sump.

_____ 7. If a sump pump is installed, provisions must be made for backup
_____.
A. drain
B. electrical power
C. auxiliary drain
D. None of the above

_____   8. Pumps that can handle sanitary sewage are referred as _____
_____   pumps, while _____ pumps are used to lift clear water.

_____   9. A foundation drain collects groundwater and channels it into a
                                       basin called a(n) _____.
                                       A. sump
                                       B. tank
                                       C. pipe
                                       D. evaporator

_____   10. A sump pump is turned on automatically by a(n) _____-controlled
                                       switch.

11. Why should a check valve be installed in the vertical pipe?

   _____

   _____

_____   12. The radioactive element _____ decomposes below the earth's
                                       surface and releases radon gas.

_____   13. Radon gas can enter the basement of a building through the foun-
                                       dation drain piping and the sump, posing a health hazard.
                                       A. True
                                       B. False

_____   14. If radon is a problem in the area, install a(n) _____ sump to expel
                                       the radon gas to the outside of the building.
                                       A. sewage
                                       B. vented
                                       C. reducing
                                       D. one-pipe

_____   15. Exterior stairwells, window wells, and ground or driveways that
                                       _____ toward the building all pose storm drainage problems.

_____   16. A(n) _____ drain should be installed in driveways that slope
                                       toward the building.

_____   17. A(n) _____ drain collects and drains water away from the founda-
                                       tion.

_____   18. The procedures for installing storm drainage piping are similar to
                                       those described for the installation of horizontal DWV piping
                                       below the concrete floor.
                                       A. True
                                       B. False

_____   19. _____ drains are installed alongside the footings inside the
                                       building.
                                       A. Foundation
                                       B. Hub
                                       C. Trench
                                       D. French

Name _____

_____ 20. Leaders from downspouts do *not* require a minimum slope.
  A. True
  B. False

_____ 21. Horizontal runs of storm water piping must maintain a minimum slope of 1/8″ per foot.
  A. True
  B. False

_____ 22. Trench drains use _____ to provide an area that will absorb water quickly.

23. What may happen to the gravel in trench drains over time?

_____

_____

Installing HVAC Systems

25

Objective: You will be able to describe the basic operation of each HVAC system and its parts and explain the jobs plumbers are required to do while installing the systems.

Instructions: *Carefully read Chapter 25 in the text and answer the following questions.*

_____ 1. A _____ provides moisture controls in the air-conditioning system.
   A. dehumidifier
   B. refrigeration unit
   C. humidifier
   D. Both A and C.

_____ 2. Heat is transferred in _____ direction(s).
   A. one
   B. two
   C. three
   D. four

_____ 3. Heat moves from a cooler object to a warmer object.
   A. True
   B. False

_____ 4. Gas and electricity are the only two energy sources that produce heat.
   A. True
   B. False

_____ 5. _____ has almost completely replaced individual area heaters.

_____ 6. In perimeter heating, the heat outlets should be installed on the _____ walls of a structure.

_____ 7. In a hydronic heating system, water is heated by a(n) _____.
   A. furnace
   B. boiler
   C. electric baseboard
   D. None of the above.

_____   8. A _____ valve is installed as a safety apparatus on the hot water
boiler.
A. gate
B. relief
C. boiler
D. make up

_____   9. A (n) _____ valve limits the pressure of the incoming water supply.

_____   10. Air vents in a hydronic heating system allow any air that enters
the system to escape at high points in the piping.
A. True
B. False

_____   11. A single-pipe hydronic system heats a building more uniformly.
A. True
B. False

_____   12. The two basic cooling systems in common use are the _____
_____   system and the _____ system.

_____   13. In a cooling system, heat from the air is absorbed by a liquid
refrigerant circulating inside the _____.
A. condenser
B. heat exchange coil
C. evaporator
D. starter

_____   14. A(n) _____ valve is used with a heat pump to produce heated air
on the heating cycle and cooled air on the cooling cycle.

15. Define the term _relative humidity_.

_____

_____

_____

_____   16. The comfort range for most people is _____ relative humidity.
A. 20% to 60%
B. 30% to 50%
C. 30% to 70%
D. 90% to 100%

_____   17. The process of changing air within an enclosed space by supplying
and distributing fresh air and exhausting used air is called _____.

_____   18. Generally, the plumber's role in the installation of an HVAC
system is limited to installation of piping to provide fuel.
A. True
B. False

_____   19. Gas-fired heating units require piping of natural gas from the
_____ to the heating unit.

Name _____

_____ 20. The purpose of a(n) _____ is to catch any foreign particles in the gas piping that might damage the heating unit and valves.

21. Match the burner unit parts with the proper name.

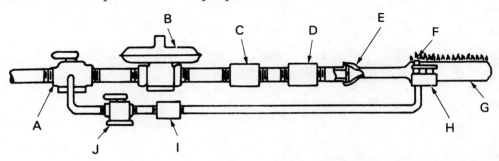

North American Heating and Air Conditioning Wholesalers Assoc.

_____ a. Safety shutoff valve

_____ b. Automatic main line valve

_____ c. Venturi orifice

_____ d. Pilot burner

_____ e. B cock

_____ f. Pressure regulator

_____ g. Burner

_____ h. Pilot filter

_____ i. Pilot generator

_____ j. A cock

_____ 22. Solar heating units require installation of a storage tank for fuel oil.
A. True
B. False

_____ 23. In forced-circulation hydronic systems, it is *not* necessary to pitch horizontal sections of piping.
A. True
B. False

_____ 24. Drain valves in a hydronic system should be installed at the _____ points in the system.

_____ 25. _____ are installed to reduce vibration and prevent sagging of pipes.

Name _____ Date _____

Period _____ Course _____

Score _____

Swimming Pools, Hot Tubs, and Spas

26

Objective: You will be able to describe the components, the design considerations, and the installation of spas, hot tubs, and swimming pools.

Instructions: *Carefully read Chapter 26 in the text and answer the following questions.*

_____ 1. The basic components of spas, hot tubs, and swimming pools are similar.
  A. True
  B. False

_____ 2. _____ is a sand-concrete mixture that is applied by spraying the mixture over steel reinforced material.

_____ 3. Hot tubs are usually constructed with round or oval wooden tubs.
  A. True
  B. False

_____ 4. Many spas and hot tubs do *not* require direct plumbing and are commonly filled with a hose.
  A. True
  B. False

_____ 5. The capacity of a hot tub or pool is measured in _____.

_____ 6. The _____ is the maximum number of people who will use the facility per hour.

_____ 7. The _____ is the frequency with which the total volume of water in the tub or pool is circulated through the filter.

_____ 8. Most public pools operate with a(n) _____ hour turnover rate so they can accommodate a larger number of people.
  A. 1–3
  B. 3–5
  C. 8–11
  D. 6–8

_____ 9. _____ remove contaminants that would otherwise make the water unsuitable for bathing.

_____ 10. _____ is the process of running water through the filter in the
opposite direction.
A. Filtering
B. Backwashing
C. Back siphoning
D. Pumped filtration

_____ 11. _____ filters consist of a series of layers of porous material that is
covered with diatomaceous earth.
A. Sand
B. DE
C. Cartridge
D. Backwash

_____ 12. _____ are used to circulate the water through filters and the heater.

_____ 13. A spa outfitted with hydrojets requires a smaller pump.
A. True
B. False

_____ 14. The design of a new installation of a large pool needs to be
approved by the local health department.
A. True
B. False

_____ 15. Spas and hot tubs usually do _not_ have heaters.
A. True
B. False

_____ 16. Floating debris is removed by _____.
A. scum gutters
B. sand filters
C. skimmers
D. Both A and C.

_____ 17. _____ give spas and hot tubs the bubbling effect.
A. Pumps
B. Valves
C. Pipes
D. Blowers

_____ 18. Swimming pools may have _____ to automatically add chlorine to
the water for purification purposes.

_____ 19. Hydrostatic pressure may cause damage to the piping when a
pool is empty.
A. True
B. False

_____ 20. A(n) _____ is installed on the inlet side of the pump to remove
hair and debris that would otherwise damage the pump.

_____ 21. Plans furnished by the engineer should be followed exactly to
_____ prevent _____ or _____ pipe and fittings.

Name _____ Date_____

Period_____ Course _____

Score_____

Irrigation Systems

Objective: You will be able to list four basic considerations for operation of sprinkler irrigation systems, explain the importance of water pressure, and list factors that can cause water pressure loss. You will be able to describe the processes of designing and installing a lawn or garden sprinkler irrigation system and describe the differences between a sprinkler irrigation system and a drip irrigation system.

Instructions: *Carefully read Chapter 27 in the text and answer the following questions.*

_____ 1. Lawn or garden sprinkler irrigation systems consist of an above ground network of piping and sprinkler heads.
  A. True
  B. False

2. List the three basic design considerations for a sprinkler irrigation system.

_____

_____

_____

_____

3. List three factors that must be considered when deciding on the type of sprinkler head to use.

_____

_____

_____

_____ 4. _____ is the resistance to water flow exerted by the walls of the pipe and fittings.
  A. Flow resistance
  B. Friction
  C. Low pressure
  D. PSIC

_____ 5. The designer can minimize pressure drop by using long pipe runs with numerous fittings.
   A. True
   B. False

_____ 6. Drains should be placed at _____ points in the sprinkler irrigation system.
   A. low
   B. high
   C. five
   D. no

7. List the three principle categories of sprinkler heads.

_____

_____

_____

_____ 8. The _____-type sprinkler head is probably the most common type for residential lawn watering.

_____ 9. The _____ spray head does *not* have a pop-up feature and is generally installed in flowerbeds.

_____ 10. _____ sprinkler heads are designed to cover a large rectangular area.

_____ 11. All sprinkler irrigation systems are installed using exactly the same installation requirements and procedure.
   A. True
   B. False

_____ 12. Working with the designer's plot plan, the installer will need to _____.
   A. locate the water source
   B. find a location for the controls
   C. lay out the sprinkler heads
   D. All of the above.

_____ 13. A(n) _____ cuts through the ground and buries the pipe all in a single operation.
   A. automatic pipe-laying machine
   B. trencher
   C. backhoe
   D. square-tipped shovel

14. Why is polyethylene preferred over polyvinyl chloride pipe?

_____

_____

Name _____

_____ 15. All pipes of a sprinkler irrigation system must slope uniformly toward one of the drain valves.
A. True
B. False

_____ 16. A sprinkler irrigation system is generally connected to existing hose bibbs by a faucet adapter.
A. True
B. False

_____ 17. A _____ system avoids much of the evaporation loss common to sprinkler irrigation systems that spray water into the air.
A. drip irrigation
B. flow restricted
C. rotary-type
D. spray-type

_____ 18. The flow of a drip irrigation system is controlled by _____ that emit 1/2 to 2 gallons of water per hour.

_____ 19. Some drip irrigation systems include a porous _____ that may be used to water rows of closely spaced plants.

20. Match each of the parts of the drip irrigation connection with its correct name.

_____ a. Tubing

_____ b. Filter

_____ c. Pressure regulator

_____ d. Faucet or hose

_____ e. Vacuum breaker

_____ f. Tubing adapter

Drip Irrigation Connection

Wade Mfg. Co.

Repairing DWV Systems

28

Objective: You will be able to recognize DWV system problems, describe methods of checking and testing a plumbing system, and explain procedures for making proper plumbing repairs.

Instructions: *Carefully read Chapter 28 in the text and answer the following questions.*

1. Name the three major groups of DWV problems.

   _____

   _____

   _____

_____ 2. No repair should be made without first isolating the problem.
   A. True
   B. False

3. Troubleshooting and repair should be done by following a three-step process. List these steps.

   _____

   _____

   _____

_____ 4. DWV piping repairs do *not* have to conform to local code.
   A. True
   B. False

5. List the two common drainage malfunctions of a toilet.

   _____

   _____

_____ 6. The simplest way to clear a blocked toilet is to use _____.
A. a snake
B. a plunger
C. air
D. a cable

_____ 7. A water ram uses air to apply high pressure to force the blockage
on through the pipe.
A. True
B. False

_____ 8. A _____ should be used if the force cup and water ram fail to
remove the obstruction.
A. closet auger
B. plunger
C. snake
D. canister auger

_____ 9. Before removing a toilet, turn off the water supply and flush the
toilet to _____ the water from the tank.
A. fill
B. empty
C. clean
D. replace

_____ 10. When replacing a toilet, it is strongly recommended to replace the
_____ to prevent a possible leak at the base of the bowl.
A. closet flange
B. closet seat
C. float ball
D. wax ring

11. Identify the parts involved in removing a toilet.

_____ a. Closet bolts

_____ b. Washer

_____ c. Wall flange

_____ d. Slip joint nut

_____ e. Nut

_____ f. Cap

_____ g. Rubber washer

_____ h. Flexible supply

_____ i. Valve

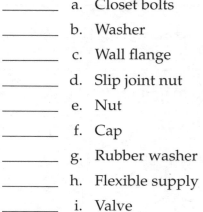
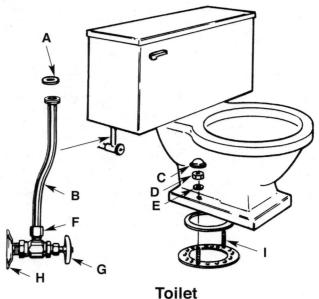

Toilet

Goodheart-Willcox Publisher

Name _____

_____ 12. It is possible to look inside DWV piping for blockages using _____ equipment

_____ 13. A blocked stack should never be cleared by running a snake or sewer machine down the vent stack.
A. True
B. False

14. List three causes of leaks near the base of a toilet bowl.

_____

_____

_____

15. List five reasons for improper flow of a tub, lavatory, shower, sink, or bidet drain.

_____

_____

_____

_____

_____

_____ 16. If a stack is clear and the water still has not drained out of the fixture, the problem is most likely in the _____.

_____ 17. A DWV pipe drain may be blocked by grease, soap, and _____.

_____ 18. If a P-trap is fitted with a(n) _____, it is only necessary to remove the plug to inspect the condition of the trap.

19. Match each of the parts of the P-trap assembly with its proper part name.

_____ a. P-trap

_____ b. Union nut

_____ c. Wall flange

_____ d. DWV branch piping

_____ e. Slip joint nut

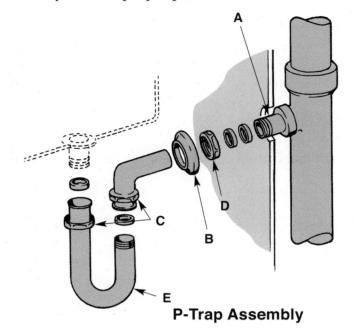

P-Trap Assembly

Goodheart-Willcox Publisher

Name _____ Date _____

Period _____ Course _____

Score _____

Repairing Water Supply Systems

Objective: You will be able to recognize water supply problems, describe methods of checking and testing a water supply system, and explain procedures for making plumbing repairs.

Instructions: *Carefully read Chapter 29 in the text and answer the following questions.*

1. List the four major groups of problems within the water supply system.

_____

_____

_____

_____

_____ 2. Faucets are often a trouble spot and need repair due to _____.
   A. weak material
   B. heavy use
   C. water pressure
   D. lack of potable water

_____ 3. A(n) _____ faucet uses a rubberlike washer that is squeezed against a seat to shut of the flow of water.

_____ 4. A valve or faucet that does *not* completely stop water flow may be caused by a _____.
   A. worn or pitted seat
   B. deteriorated or broken washer
   C. defective handle or stem
   D. All of the above.

_____ 5. If valve seats are pitted or worn, they must be _____.
   A. threaded
   B. refaced
   C. replaced
   D. Both A and B.

_____ 6. The faucet handle should rotate without changing water flow.
   A. True
   B. False

7. Match each part of the faucet with its correct name.

_____ a. Handle

_____ b. Packing

_____ c. Spline

_____ d. Bonnet

_____ e. Stem

_____ f. Spout

_____ g. Internal thread

_____ h. Faucet base

_____ i. Seat

_____ j. Brass screw

_____ k. Washer

_____ l. Stem threads

Faucet

Goodheart-Willcox Publisher

_____ 8. Water leakage around the stem when the faucet or valve is opened is caused by deteriorated packing or a(n) _____.

_____ 9. When water flow from a spout is slow, the shutoff valve in the water supply piping should be checked first.
   A. True
   B. False

10. Match each part of the rotating ball faucet with its correct name.

_____ a. Cam housing

_____ b. Aerator

_____ c. Spout

_____ d. Cap

_____ e. Set screw

_____ f. Ball assembly

_____ g. Cam rubber

_____ h. Diverter unit

_____ i. Stainless steel spring

_____ j. Seat washer

_____ k. O-ring washer

_____ l. O-ring

Rotating Ball Faucet

Goodheart-Willcox Publisher

Name _____

_____ 11. In a tank-type toilet, a faulty _____ will cause water to flow
continuously.
A. overflow tube
B. float valve
C. float ball
D. All of the above.

_____ 12. The two types of pressure flush valves are diaphragm and _____.
A. piston
B. flapper
C. ball
D. float

13. Match each part of the diaphragm-type pressure valve with its correct name.

_____ a. Auxiliary valve

_____ b. Plunger

_____ c. Handle

_____ d. Lower chamber

_____ e. Upper chamber

_____ f. Diaphragm

_____ g. Bypass

_____ h. Inlet

_____ i. Outlet

Diaphragm-Type Pressure Valve

Sloan Valve Co.

_____ 14. When water is leaking from the toilet tank, a _____ is a possible
cause.
A. cracked tank
B. loose inlet water supply
C. condensation on the outside of the tank
D. All of the above.

_____ 15. The spud washer is located at the joint of the tank and bowl.
A. True
B. False

_____ 16. If a float ball does *not* rise to the surface, it is leaking and must be
_____.

_____ 17. If the ball or flapper on a flush valve is worn, it will *not* seal
properly.
A. True
B. False

_____
_____ 18. Pressure flush valves are of two types, _____ and _____.

_____ 19. In a diaphragm pressure flush valve, the _____ upsets the relief
valve and causes water to flow past the diaphragm to the fixture.
A. plunger
B. vacuum breaker
C. water stop
D. closet trap

20. The maximum consumption of water for each of the following fixtures, as set forth by the
National Plumbing Standards Act passed by Congress in 1992, is:

Showerheads _____

Faucets _____

Toilets _____

21. List five problems that are likely to occur with the water supply system.

_____

_____

_____

22. What are the three principal causes of leaks in piping?

_____

_____

_____ 23. The term used to describe the banging noise that can result from
the sudden stopping of water flow in a piping system is _____.

24. List four ways frozen metal pipe may be thawed.

_____

_____

_____ 25. Plumbers should avoid using a torch to thaw frozen pipes.
A. True
B. False

Name _____

_____ 26. Plumbers must be sure the building water line is buried below the
_____ line.
   A. sewer
   B. vent
   C. wall
   D. frost

_____ 27. A water heater tank must be drained periodically to remove the
_____ that collects at the bottom.

_____ 28. A few simple maintenance procedures can prolong the life of a
water heater and increase operating safety.
   A. True
   B. False

_____ 29. A water heater tank should be drained every 18 months to remove
sediment in the tank.
   A. True
   B. False

30. Using the numbers 1 through 6, indicate the proper order of the steps to ignite the pilot on a
water heater.

_____ a. Release the reset button.

_____ b. Depress and hold reset button.

_____ c. Replace cover and turn thermostat to desired temperature.

_____ d. Set thermostat valve on pilot.

_____ e. Hold reset button down for one minute or until thermocouple is heated.

_____ f. Hold a lighted wooden match or charcoal lighter in front of pilot orifice.

_____ 31. When an electric water heater does not produce hot water, check
the _____ first.

_____ 32. The most obvious cause for shortage of hot water is a heater that
is too large.
   A. True
   B. False

_____ 33. If the water is too hot, the _____ is the most likely cause.
   A. thermocouple
   B. energy source
   C. thermostat
   D. dip tube

_____ 34. An electric water heater should *not* be turned on before the tank is
filled.
   A. True
   B. False

_____   35. A rumbling sound in a hot water heater tank as the water is
heating is probably caused by _____ in the tank.
A. air
B. sediment
C. cold water
D. leakage

_____   36. The most common cause of freezing of water pipes is _____.
A. thin walled pipe
B. abnormally low temperatures
C. improper installation
D. Both B and C.

_____   37. A leak in the _____ of a hot water heater may allow cold water to
mix without outgoing hot water.

Name _____ Date _____
Period_____ Course _____
Score_____

Remodeling

30

Objective: You will be able to manage a plumbing remodeling job, replace fixtures and faucets without causing damage to other items, complete a roughed-in bathroom, relocate existing plumbing fixtures, determine if fixtures can be added to existing piping, and install new plumbing in an addition to an existing building.

Instructions: *Carefully read Chapter 30 in the text and answer the following questions.*

_____ 1. In a remodeling job, existing piping refers to the piping that was installed during the _____-rough stage when the building was first constructed.

2. What steps can be taken to prevent the floor during remodeling?

_____

_____

_____

_____

_____ 3. Heating and air-conditioning return air ducts in the work area should be blocked to prevent _____ from entering.

_____ 4. The amount of modification to the DWV piping can greatly
_____ increase the _____ and _____ of completing a remodeling project.

_____ 5. Remodeling work is most often done while the building is unoccupied.
A. True
B. False

_____ 6. Code requirements must be considered during the planning and performing of a remodeling project.
A. True
B. False

_____ 7. The shortest route is always the best route for workers to get from the entrance to the work area.
   A. True
   B. False

_____ 8. The plumber is responsible for ensuring that the structure is protected while performing the plumbing remodeling.
   A. True
   B. False

_____ 9. The first step when replacing a faucet is _____.
   A. disconnect water supplies
   B. remove P-trap
   C. turn off water
   D. None of the above.

_____ 10. The second step when replacing a faucet is _____.
   A. turn off water
   B. remove P-trap
   C. replace water supply
   D. disconnect the water supply

_____ 11. If the P-trap is removed, the DWV stub-out should be plugged to prevent _____ from entering the room.
   A. propane gas
   B. sewer gas
   C. water condensation
   D. natural gas

_____ 12. Plumbing codes require that the DWV piping serving a water closet be a minimum of _____.
   A. 2″
   B. 3″
   C. 4″
   D. 5″

_____ 13. Removing an enameled cast-iron tub is difficult because of the
_____   fixture's _____ and _____.

14. What safety items should be worn by a plumber when using a sledge hammer to break up an enameled cast-iron tub?

_____

_____

_____ 15. The distance from the lavatory trap to the vent for the lavatory is typically limited by code.
   A. True
   B. False

_____ 16. Developed length is measured along the _____ of the pipe and fittings.

Name _____

_____ 17. When relocating fixtures, the first question that needs to be
answered is, "Where are the _____ pipes located?"
   A. lavatory
   B. DWV
   C. gas
   D. water

_____ 18. If the DWV piping is modified to change the location of a fixture,
then the _____ piping will need to be changed also.

_____ 19. When adding new fixtures, the _____ of the existing DWV piping
must be evaluated to determine if it can carry the additional load.

_____ 20. Code requires that cleanouts are required at each change of direc-
tion that is greater than _____°.
   A. 45
   B. 22-1/2
   C. 60
   D. 11-1/2

_____ 21. When remodeling requires the installation of new fixtures, the
_____ of the water supply pipe must be considered.

22. What does the abbreviation *dfu* stand for?

_____

23. What does the abbreviation *wsfu* stand for?

_____

_____ 24. The maximum number of dfu on a 4″ horizontal sewer pipe with
a slope of 1/8″ per foot of run is _____.

_____ 25. The maximum number of dfu on a 4″ horizontal sewer pipe with
a slope of 1/4″ per foot of run is _____.

Job Organization

31

Objective: You will be able to describe how to become familiar with a new plumbing job, identify the materials and equipment necessary to complete a plumbing job, make better use of your time, describe the importance of teamwork on the job, and describe how to coordinate plumbing with the work of other trades.

Instructions: *Carefully read Chapter 31 in the text and answer the following questions.*

_____ 1. For a new building, the _____ should be reviewed in order to understand the requirements of the job.
   A. drawings
   B. specifications
   C. plans
   D. All of the above.

_____ 2. A(n) _____ is used to identify plumbing requirements and contains questions that can be answered by reviewing plans and specifications.

3. How should a plumber determine the size of pipe if plumbing drawings are not included for a residential structure?

_____

_____ 4. The building water and sewer lines are installed during the _____ stage.
   A. finish
   B. first-rough
   C. second-rough
   D. Either B or C.

_____ 5. All pipe and fittings that will be covered in the finished structure are installed during the _____ stage.
   A. first-rough
   B. second-rough
   C. third-rough
   D. Either A or C.

_____   6. All materials to complete a plumbing job should be delivered at one time and dropped off in one place.
    A. True
    B. False

_____   7. The installation of fixtures, faucets, appliances, and water heaters are performed during the _____ stage.
    A. rough
    B. finish
    C. first-rough
    D. second-rough

_____   8. The _____ should indicate the materials, fixtures, and faucets to be used.
    A. specifications
    B. plumbing code
    C. mechanical code
    D. plumbing instructor

_____   9. When adding a bathroom to an existing building, the first-rough stage is eliminated.
    A. True
    B. False

_____   10. When preparing a list of plumbing materials and supplies, the quantity, size, and _____ of each item should be indicated.

_____   11. The list of materials should be divided into materials for the various stages of work.
    A. True
    B. False

_____   12. The number of people in the plumbing crew and their skills must be considered in planning.
    A. True
    B. False

_____   13. It is good practice to have slightly more materials and supplies available than is estimated to be installed in _____ of work.
    A. one day
    B. two days
    C. three days
    D. one week

_____   14. The plumber is responsible for ensuring that all tools are in safe and working condition.
    A. True
    B. False

_____   15. During construction, the most obvious need for coordination of the different trades relates to _____.
    A. payment
    B. scheduling
    C. equipment availability
    D. None of the above.

Name _____

16. In addition to developing an environment that encourages teamwork what are three keys to teamwork?

_____

_____

_____

_____

_____

_____ 17. To develop an environment that encourages teamwork, employers need to emphasize the need for _____ among the members of the work group.

_____ 18. Learning to work smarter primarily relates to thinking _____.

19. List two ways plumbers can be prepared to respond to injuries if they occur.

_____

_____

_____

_____

_____ 20. Generally, the _____ is expected to furnish all of the equipment necessary to perform an inspection.

_____ 21. To be a skilled plumber, it is necessary to achieve both quality and a reasonable level of speed in performing tasks.
A. True
B. False

Plumbing Career Opportunities

32

Objective: You will be able to identify sources of plumbing jobs, explain the levels of the plumbing apprenticeship program and the program's educational requirements, and list qualifications for success in the plumbing trade. You will be able to describe characteristics of desirable employees, describe the process of getting and keeping a job, and explain the process of starting a small business.

Instructions: *Carefully read Chapter 32 in the text and answer the following questions.*

1. List five related areas in which a plumber might find employment.

   _____

   _____

   _____

_____ 2. Plumbing _____ serve the needs of the community and offer the greatest opportunity as an entry point for most beginners.

_____ 3. Plumbing enterprises include retailers as well as plumbing installation and repair services.
   A. True
   B. False

4. Briefly list four functions of government agencies that hire plumbers.

   _____

   _____

   _____

   _____

_____ 5. Plumbing educators generally are master plumbers with many years of experience.
   A. True
   B. False

_____ 6. Plumbing _____ provide benefits to the plumber, and are the source of employees for numerous contractors.

_____ 7. An apprentice plumber goes through an educational program that is generally _____ years long.

_____ 8. Journeyman plumbers are qualified to be plumbing contractors.
  A. True
  B. False

_____ 9. Master plumbers are generally required to work for _____ years before taking the exam for a master's license.

_____ 10. _____ are responsible for directing the work of a small group of workers.

_____ 11. _____ oversee large plumbing jobs and generally have several supervisors working under their direction.

_____ 12. _____ are responsible for estimating the material and actual cost of a job.

_____ 13. Plumbing _____ are master plumbers who generally hire apprentices, journeymen, and other master plumbers to work for them.

14. What is the most important function of a plumbing supply dealer?

_____

_____

_____ 15. One of the most important skills that a plumber must acquire is the ability to visualize the completed plumbing system before work is started.
  A. True
  B. False

_____ 16. Only _____ plumbers are eligible to obtain plumbing permits.

_____ 17. The three-part foundation of basic skills, _____, and _____ can be considered fundamental to everyone's education.
_____

18. List the five skills essential for success in the workplace.

_____

_____

_____ 19. The mathematics-related skill that is most frequently used by plumbers is _____.

Name _____

_____ 20. Receiving, attending to, interpreting, and responding to verbal messages are all a part of _____ skills.
  A. speaking
  B. writing
  C. listening
  D. caring

21. List two types of speech to avoid in the workplace.

_____

_____

22. List the six thinking skills identified by the Labor Department's report.

_____

_____

_____

_____ 23. Which of the following is *not* a personal quality identified by the Labor Department?
  A. integrity
  B. sociability
  C. efficiency
  D. responsibility

24. What three things should a plumber do if he or she makes a mistake on the job?

_____

_____

_____

_____ 25. A plumber's confidence in his or her ability enables the plumber
_____ to proceed with the work in a(n) _____ and _____ manner.

_____ 26. A plumber who has the ability to get along with diverse groups of people exhibits the personal quality of _____.

_____ 27. Self-_____ involves exercising self-control or self-discipline.

_____ 28. Integrity involves being _____.
  A. honest
  B. trustworthy
  C. dependable
  D. All of the above.

29. List five resources a plumber must effectively manage.

_____

_____

_____ 30. Effectively using tools and equipment begins with selecting the appropriate tool for the task.
A. True
B. False

31. Match each of the following words with the statement that describes it.

_____ a. Punctual   A. Gives a fair day's work

_____ b. Dependable   B. Works willingly with other plumbers

_____ c. Honesty   C. Arrives on time and ready to work

_____ d. Motivation   D. Follows through on work assignments

_____ e. Technical ability E. Work consistently meets employer's standard

_____ f. Leadership   F. Truthful with the supervisor and coworkers

_____ g. Quality of work   G. Guides others toward the completion of assigned work

_____ h. Cooperative   H. Performs the work with limited supervision

_____ 32. A family member is an example of a good reference.
A. True
B. False

_____ 33. Finding employment opportunities through other people, such as friends, is called _____.

_____ 34. Plumbing suppliers are a source of job leads.
A. True
B. False

_____ 35. A plumber's _____ provides documentation of work experience and may contain plans, sketches, photographs, and a list of jobs the plumber has completed.

Name _____

36. How is becoming a part-time, self-employed plumber helpful before starting a business?

_____

_____

_____

_____

_____

_____

37. What two people must the owner of a small business work with in order to prepare a business plan?

_____

Plumbing History

33

Objective: You will be able to describe examples of plumbing developments from ancient times; identify contributions made by the Romans; trace the development of plumbing in Western Europe and the United States; and describe the development of plumbing fixtures, piping materials, and tools. You will be able to describe the contribution of laws, codes, and organizations to the development of plumbing.

Instructions: *Carefully read Chapter 33 in the text and answer the following questions.*

_____ 1. The history of plumbing can be traced back _____ years to the Babylonians.

_____ 2. The ancient Romans were famous for their _____ and public _____.
_____

_____ 3. In Western Europe, sewers were constructed in some areas to help control _____ water.

_____ 4. London's cholera epidemic in the early 1800s was traced to the water supply.
A. True
B. False

5. When did plumbing as we know it in the United States begin to develop?

_____

_____ 6. The _____ Empire was the first to provide large numbers of its citizens with plumbing.
A. Greek
B. Roman
C. Babylonian
D. British

_____ 7. The Black Plague was directly connected to the lack of sanitary waste disposal.
A. True
B. False

8. Match the following terms and descriptive phrases.

_____ a. Disease traced to contaminated drinking water   A. Dark Ages

_____ b. 400–1000 CE   B. Black Plague

_____ c. Caused the death of 25 million people   C. Cholera

_____ 9. A(n) _____ epidemic in 1854 killed five percent of Chicago's population.

10. Why was Chicago's water supply contaminated in the mid-1800s?

_____

_____

_____

_____

_____

_____ 11. A New York state law passed in 1881 required that plumbers be
_____.
A. trained
B. registered
C. union members
D. None of the above.

_____ 12. Early attempts to install plumbing indoors failed because _____
_____ escaped into the house even though _____ were installed.

_____ 13. The general idea of _____ drain/waste piping systems was first
used in New York in 1874.
A. building
B. cleaning
C. venting
D. None of the above.

_____ 14. What size pipe was first used to vent traps in New York City?
A. 1/2″
B. 1″
C. 1-1/2″
D. 2″

_____ 15. The Greeks' major contribution to plumbing was the development
of bathtubs with _____.

Name _____

_____ 16. Bitumen is a natural type of _____ that was used to make clay floors waterproof in about 3000 BCE.
   A. clay
   B. asphalt
   C. metal
   D. polymers

_____ 17. The Romans used lead to make pipe and line sinks and other fixtures.
   A. True
   B. False

_____ 18. Plumbum is the Latin word for _____.

_____ 19. In 1562, _____ was used to supply water to a fountain in Germany.

_____ 20. Early bathtubs were made of _____ in the shape of a(n) _____.
_____

_____ 21. Evidence of _____ has been found in the ruins of Egyptian cities that date back to 1350 BCE.

_____ 22. In 1775, _____ of England patented the water closet that became the prototype for today's toilets.
   A. Alexander Cummings
   B. Alexander Bell
   C. Sir John Harington
   D. George Jennings

_____ 23. In 1906, _____ applied for a patent for the flush valve, which made possible the tankless toilet used in most public restrooms.
   A. John Harington
   B. Joseph Bramah
   C. William E. Sloan
   D. Thomas Twyford

_____ 24. Boston set up the first waterworks in the United States in 1652 to supply water to wharves and buildings for _____ and domestic use.

_____ 25. The automatic storage water heater was invented _____.
   A. in the 1930s
   B. by an employee of the Kohler Co.
   C. by Edwin Ruud
   D. All of the above.

_____ 26. Philadelphia became the first city in the United States to select cast iron for water mains.
   A. True
   B. False

_____ 27. Galvanized pipe is coated with _____ metal to prevent rust.

_____ 28. PVC pipe began to be manufactured during the 19th century.
  A. True
  B. False

_____ 29. The first compact trencher was called the _____.

30. In which year were the following federal laws enacted?

_____ a. Occupational Safety and Health Act

_____ b. Safe Drinking Water Act

_____ c. Americans with Disabilities Act

_____ 31. Water efficiency regulations for plumbing fixtures were established by the _____.
  A. Hoover Code
  B. International Residential Code for One- and Two-Family Dwellings
  C. Uniform Plumbing Code
  D. National Plumbing Standard

Name _____ Date_____

Period_____ Course _____

Score_____

Safety

**Job
1**

Text Reference

Chapter 1, Pages 3–34

Objective

After completing this job, the student will have demonstrated the ability to explain the importance of safety and inspect personal protective equipment, such as hard hats, safety goggles, safety shoes, hand tools, extension cords, and power-tool cords to ensure that they are free from breaks or other damage.

Instructions

1. Explain the importance of safety and developing safe work habits.   Completed ❏

   _____

   _____

   _____

   _____

2. Inspect the safety items.   Completed ❏

   a. Hard hat

   b. Safety goggles

   c. Safety shoes

   d. Safety harness

   e. First aid kit

3. Inspect the hand tools.   Completed ❑

   a. Wood chisel for sharpness

   b. Chisel for mushroom head

   c. Screwdriver for broken handle

   d. Fire extinguisher

4. Inspect the following items.   Completed ❑

   a. Extension cord

   b. Power tool cord

   c. Stepladder

   d. Rolling scaffold

Instructor's Initials _____

Date _____

Math Calculations

Text Reference

Chapter 4, Pages 79–80

Objective

After completing this job, the student will have demonstrated the ability to solve math problems using various math formulas and equations.

Instructions

Calculate and answer the following problems using the formulas discussed in the text material.

1. Solve the area of the following rectangles.   Completed ❑

   a. L = 25′, W = 15′   Area = _____

   b. L = 65′, W = 20′   Area = _____

   c. L = 19′, W = 6′   Area = _____

   d. L = 27.5′, W = 5.5′   Area = _____

   e. L =17 yd, W = 9 yd   Area = _____

   f. L = 100′, W = 18′   Area = _____

   g. L = 40″, W = 16″   Area = _____

   h. L = 60 yd, W = 30 yd   Area = _____

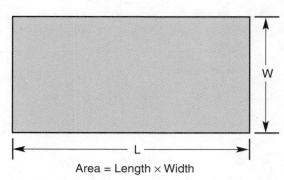

Area = Length × Width

Goodheart-Willcox Publisher

2. Solve the area of the following circles based on the information provided.

Completed ❑

a. Diameter of 16′   Area = _____

b. Radius of 8′   Area = _____

c. Diameter of 12′   Area = _____

d. Radius of 9′   Area = _____

e. Diameter of 15′   Area = _____

f. Diameter of 22′   Area = _____

g. Radius of 6′   Area = _____

h. Radius of 12′   Area = _____

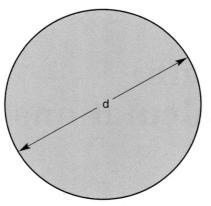

Area = πr^2
π = 3.14
r = Radius = ½ diameter

Goodheart-Willcox Publisher

3. Solve the volume of the following rectangular tanks in cubic feet.

Completed ❑

a. L = 20′, W = 20′, H = 40′   Volume = _____

b. L = 12′, W = 6′, H = 20′   Volume = _____

c. L = 28′, W = 8′, H = 8′   Volume = _____

d. L = 10′, W = 8′, H = 16′   Volume = _____

e. L = 18′, W = 8′, H = 10′   Volume = _____

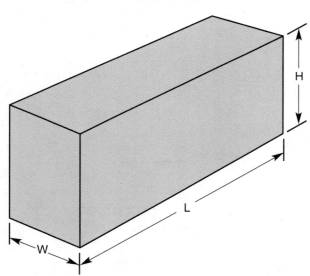

Volume = Length (L) × Width (W) × Height (H)

Goodheart-Willcox Publisher

Name _____

4. Solve the volume of the following cylindrical tanks in gallons.   Completed ❑

    a. Radius 5′, Height 15′   Volume = _____

    b. Radius 12′, Height 20′   Volume = _____

    c. Radius 3′, Height 12′   Volume = _____

    d. Diameter 10′, Height 30′   Volume = _____

    e. Diameter 20′, Height 60′   Volume = _____

    f. Radius 4′, Height 22′   Volume = _____

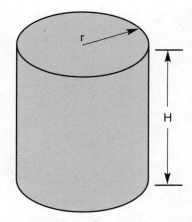

Volume = PI × radius squared × Height
This formula may be written: $V = \pi^2 h$

Goodheart-Willcox Publisher

Instructor's Initials _____

Date _____

Name _____ Date_____

Period_____ Course _____

Score_____

Job

3

Calculating Slope on a DWV Horizontal Sewer

Text Reference

Chapter 4, Pages 80–81

Objective

After completing this job, the student will have demonstrated the ability to calculate the amount of slope on a horizontal sewer.

Instructions

1. Using the following illustration, calculate the slope of a horizontal sewer to be installed.

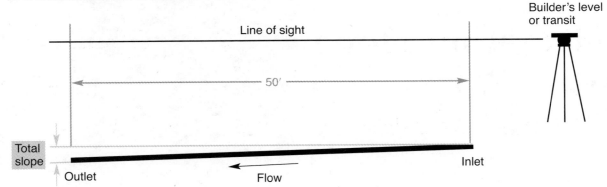

Total slope = Slope per foot × Feet of run
= ⅛″ × 50
= ⁵⁰⁄₈″
= 6¼″

Goodheart-Willcox Publisher

Slope per foot	x	**Feet of run**		**Total slope**
a. 1/8″		80	=	_____
b. 1/8″		100	=	_____
c. 1/4″		100	=	_____
d. 1/16″		100	=	_____
e. 1/2″		72	=	_____
f. 1/8″		48	=	_____
g. 1/2″		75	=	_____
h. 3/16″		100	=	_____

Instructor's Initials _____

Date _____

Blueprint Drawing

Text Reference

Chapter 6, Pages 95–110

Objective

After completing this job, the student will be able to design the drainage system for a simple bathroom and make an isometric drawing of this design.

Instructions

1. From the floor plan, design the drainage system for the bathroom.   Completed ❑

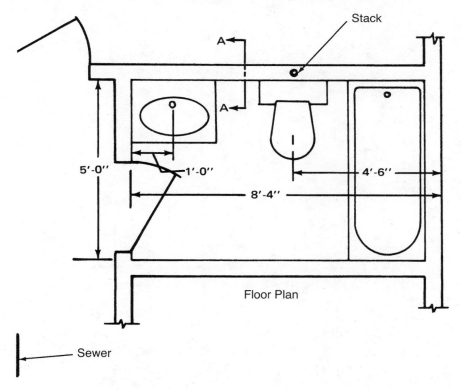

Floor Plan

Sewer

Goodheart-Willcox Publisher

2. Make an isometric drawing of the design.   Completed ❑

Instructor's Initials _____

Date _____

Job
5

Blueprint Specification Interpretation

Introduction

Blueprint reading is a very important part of the plumbing installation process. Plumbers must be able to read the blueprints in order to locate walls and pipe chases during the first and second rough-in stages. It is imperative as an apprentice to begin learning how to read and interpret the information on the blueprints. In addition, plumbers must read and interpret (understand) plumbing fixture specification sheets in order to locate where to stub out drain-pipes and water piping stub-outs for the various fixtures. This job will challenge the student to do just that.

Text Reference

Chapter 13, Pages 182–184
Chapter 22, Pages 366–367

Tools and Equipment

Lavatory Fixture Specification Sheet
Wheelchair Lavatory Specification Sheet
Round Front/Elongated Toilet Specification Sheet
Standard Bath Tub Specification Sheet

Objective

After completing this job, the student will have demonstrated the ability to read and interpret information provided on the fixture specification sheets.

Instructions

1. Based on the information provided on the specification sheet for the *American Standard Declyn Wall-Hung Lavatory #0321.026*, answer the following questions.

   Completed ❑

   a. What material was used to make the lavatory?

   b. The size of the waste at the wall is _____.

   c. The spread of the cold and hot water stub-outs is _____.

d. The flood level rim for residential is _____ above finished floor.

e. The flood level rim for ADA compliance is _____ above finished floor.

f. The waste for residential is _____ above finished floor.

g. If the flood level rim is 31″, what is the dimension to the top of the ledge back?

_____

h. The lavatory has a(n) _____ spread for mounting the faucet.

i. The cold and hot water roughs out of the wall at _____ above finished floor.

j. The center of the two openings for the lavatory wall bracket is installed how far above finished floor? Residential: _____ ADA compliance: _____

2. Based on the information provided on the specification sheet for the *American Standard Wheelchair Lavatory #9141.011*, answer the following questions.   Completed ❏

a. What material was used to make the wheelchair lavatory?

_____

b. What is the overall dimension of the lavatory?

_____

c. The faucet opening has a(n) _____ spread.

d. The flood level rim is _____ above the finished floor.

e. The waste out at the wall is _____.

f. The waste is located _____ above finished floor.

g. What is the distance from the wall to the center of the waste outlet in the lavatory?

h. The cold and hot water is roughed out at _____ above finished floor.

i. The spread of the cold and hot water stub-outs is _____.

j. The center of the faucet openings is _____ from the wall.

3. Based on the information provided on the specification sheet for the *American Standard Elongated Toilet #2399.012*, answer the following questions.   Completed ❏

a. What is the center line of the outlet of the toilet from the wall?

b. Based on the 12″ rough of the outlet, what is the distance from the back wall to the front of the toilet bowl?

c. The toilet bowl is _____ wide.

d. What is the dimension of the toilet tank lid? _____

Name _____

    e. The top of the toilet bowl is _____ mm above the finished floor.

    f. The water supply rough-in is _____ above the finished floor.

    g. From the finished floor to the top of the tank including lid is _____ .

    h. The water supply stub-out is _____ from the center line of the DWV drain.

4. Based on the information provided on the specification sheet for the *American Standard Recess Bath #2391.202,* answer the following questions.   Completed ❑

    a. The length of the tub is _____ .

    b. The width of the tub is _____ .

    c. The tub shown in the specification sheet is a _____ hand tub.

    d. What is the size of the drain outlet?

    e. The center of the drain outlet is _____ from the back (60″) side.

    f. The top of the front side of the tub is _____ above finished floor.

    g. The center line of the valves is _____ above the finished floor.

    h. The center line of the tub spout is _____ above the finished floor.

Instructor's Initials _____

Date _____

American Standard

DECLYN™
WALL-HUNG LAVATORY
VITREOUS CHINA

DECLYN WALL-HUNG LAVATORY

- Vitreous china
- Rear overflow
- Soap depression
- Faucet ledge.
  Shown with 2000.101 Ceramix faucet
  (not included)

❑ **0321.026** With wall hanger (Illustrated)
Faucet holes on 102mm (4") centers

❑ **0321.075** For concealed arms support
Faucet holes on 102mm (4") centers

Nominal Dimensions:
483 x 432mm
(19" x 17")

Bowl sizes:
362mm (14-1/4") wide,
273mm (10-3/4") front to back,
152mm (6") deep

Fixture Dimensions conform to ANSI Standard
A112.19.2

To Be Specified
❑ Color:
❑ Faucet*:
❑ Faucet Finish:
❑ Supplies:
❑ 1-1/4" Trap:
❑ Nipple:
❑ Concealed Arms Support (by others):

* See faucet section for additional models available

 ● Top of front rim mounted 864mm (34") maximum from
finished floor.
**MEETS THE AMERICAN DISABILITIES ACT GUIDELINES
AND ANSI A117.1 REQUIREMENTS FOR PEOPLE WITH
DISABILITIES**

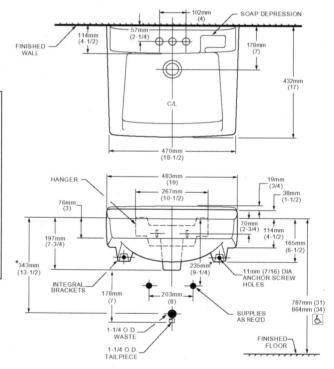

NOTES:
* DIMENSIONS SHOWN FOR LOCATION OF SUPPLIES AND "P" TRAP ARE SUGGESTED.
PROVIDE SUITABLE REINFORCEMENT FOR ALL WALL SUPPORTS.
FITTINGS NOT INCLUDED AND MUST BE ORDERED SEPARATELY.
IMPORTANT: Dimensions of fixtures are nominal and may vary within the range of tolerances
established by ANSI Standard A112.19.2.
These measurements are subject to change or cancellation. No responsibility is assumed for use
of superseded or voided pages.

SPS 0321

Revised 4/97

American Standard Inc.

American Standard

 BARRIER FREE

WHEELCHAIR USERS LAVATORY

- Vitreous china
- Front overflow
- For concealed arms support (by others)
- Faucet ledge (faucet not included)

❏ **9141.011**
    Faucet holes on 102mm (4") centers
    (Illustrated)

❏ **9141.029**
    Faucet holes on 102mm (4") centers
    • Extra right-hand hole

❏ **9141.035**
    Faucet holes on 102mm (4") centers
    • Extra left-hand hole

❏ **9141.911**
    Faucet holes on 102mm (4") centers
    • Less overflow

❏ **9140.013**
    Faucet holes on 267mm (10-1/2") centers

❏ **9140.021**
    Faucet holes on 267mm (10-1/2") centers
    • Extra right-hand hole

❏ **9140.039**
    Faucet holes on 267mm (10-1/2") centers
    • Extra left-hand hole

❏ **9140.913**
    Faucet holes on 267mm (10-1/2") centers
    • Less overflow

❏ **9140.047**
    Center hole only

❏ **9140.947**
    Center hole only
    *Less overflow

Nominal Dimensions:
508 x 686m
(20" x 27")

Compliance Certifications -
Meets or Exceeds the Following Specifications:
• ASME A112.19.2 for Vitreous China Fixture

Top of front rim mounted 864mm (34") from finished floor.
**MEETS THE AMERICAN DISABILITIES ACT GUIDE-
LINES AND ANSI A117.1 ACCESSIBLE AND USEABLE
BUILDINGS AND FACILITIES - CHECK LOCAL CODES.**

NOTE: Roughing-in information shown on reverse side of page

To Be Specified
❏   Color:
❏   Faucet*:
❏   Faucet Finish:
❏   Supplies with Stop:
❏   1-1/4" Trap:
❏   Nipple:
❏   Concealed Arms Support (by others):
❏   Offset Grid Drain Assembly: 7723.018

* See faucet section for additional models available

CI-53

© 2004 American Standard Inc.

Revised 9/04

American Standard Inc.

American Standard

♿ **BARRIER FREE**

WHEELCHAIR
USERS LAVATORY
VITREOUS CHINA

9141.011

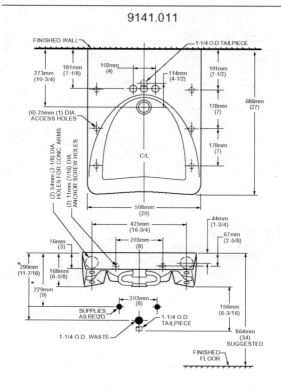

9141.029

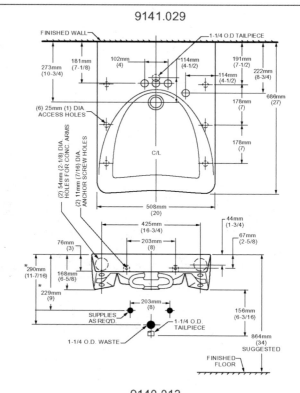

9141.035

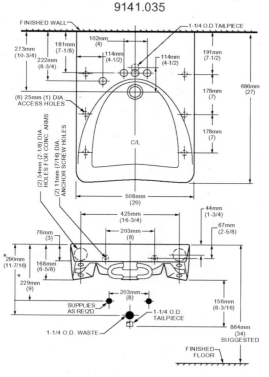

9140.013

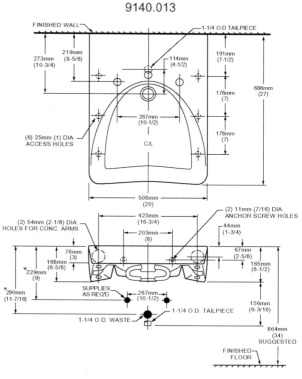

American Standard Inc.

American Standard

COLONY™
ELONGATED TOILET
VITREOUS CHINA

COLONY™ ELONGATED TOILET

☐ **2399.012**

- Vitreous china
- Low-consumption (6.0 Lpf/1.6 gpf)
- Elongated siphon action jetted bowl
- Fully glazed 2" trapway
- American Standard Aquameter™ Water Control
- Large 10" x 8" water surface area
- Color-matched trip lever
- Sanitary bar on bowl
- 2 bolt caps
- 100% factory flush tested

☐ **3344.017** Bowl

☐ **4392.016** Tank

Nominal Dimensions:
762 x 508 x 743mm (30" x 20" x 29-1/4")

Fixture only, seat and supply by others

Alternate Tank Configurations Available:

☐ **4392.500** Tank complete w/Aquaguard Liner

☐ **4392.800** Tank complete w/Trip Lever Located on Right Side

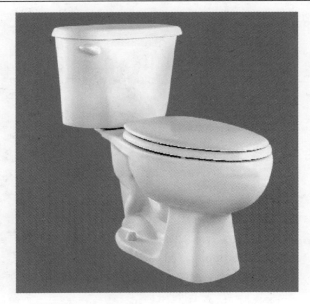

To Be Specified

☐ Color:

☐ Seat: American Standard #5324.019 "Rise and Shine" (with easy to clean lift-off hinge system) solid plastic closed front seat with cover. See pageTB-001.

☐ Seat: American Standard #5311.012 "Laurel" molded closed front seat with cover. See pageTB-001.

☐ Alternate Seat:

☐ Supply with stop:

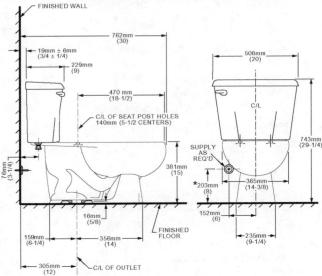

NOTES:
THIS COMBINATION IS DESIGNED TO ROUGH-IN AT A MINIMUM DIMENSION OF 305MM (12") FROM FINISHED WALL TO C/L OF OUTLET.
★ DIMENSION SHOWN FOR LOCATION OF SUPPLY IS SUGGESTED.
SUPPLY NOT INCLUDED WITH FIXTURE AND MUST BE ORDERED SEPARATELY.
IMPORTANT: Dimensions of fixtures are nominal and may vary within the range of tolerance established by ANSI Standard A112.19.2
These measurements are subject to change or cancellation. No responsibility is assumed for use of superseded or voided pages.

Compliance Certifications -
Meets or Exceeds the Following Specifications:

- ASME A112.19.2M (and 19.6M) for Vitreous China Fixtures - includes Flush Performance, Ball Pass Diameter, Trap Seal Depth and all Dimensions

SPS 2399.012

American Standard

PRINCETON™ RECESS BATH
AMERICAST® BRAND ENGINEERED MATERIAL

PRINCETON RECESS BATH
Americast® brand engineered material

- ❏ **2391.202** Right Hand Outlet
- ❏ **2391.202TC** Same as above w/tub cover
- ❏ **2390.202** Left Hand Outlet
- ❏ **2390.202TC** Same as above w/tub cover
- • Acid resistant porcelain finish
- • Recess bath with integral apron and tiling flange
- • Integral lumbar support
- • Beveled headrest
- • Full slip-resistant coverage
- • End drain outlet

PRINCETON RECESS BATH for Above Floor Rough Installation

- ❏ **2392.202** Left Hand Outlet for above floor installation
- ❏ **2392.202TC** Same as above w/tub cover
- ❏ **2393.202** Right Hand Outlet for above floor installation
- ❏ **2393.202TC** Same as above w/tub cover

Nominal Dimensions: 1524 x 762 x 356mm (445mm for above floor installation)
60" x 30" x 14" (17-1/2" for above floor installation)

Bathing Well Dimensions: 1423 x 635 x 337mm (56" x 25" x 13-1/4")

Americast® brand engineered material is a composition of porcelain bonded to enameling grade metal, bonded to a patented structural composite.

Compliance Certifications -
Meets or Exceeds the Following Specifications:
• ASME A112.19.4 for Americast Plumbing Fixtures
• ASTM F-462 for Slip-resistant Bathing Facilities
• ANSI Z124.1 Ignition Test
• ASTM E162 for Flammability
• NFPA 258 for Smoke Density

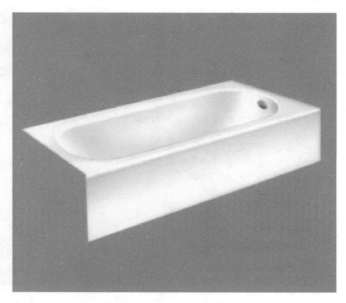

NOTE: Roughing-in dimensions shown on reverse side of page.

To Be Specified	
❏	Color:
❏	Bath Faucet*:
❏	Faucet Finish:
❏	Drain:
❏	Drain Finish:

* See faucet section for additional models available

BV-38

Revised 5/98   © 2003 American Standard Inc.

American Standard Inc.

PRINCETON™
RECESS BATH
AMERICAST® BRAND ENGINEERED MATERIAL

NOTE: Roughing-in shown for 2390/2391 only. Refer to installation instruction for 2392/2393 above-floor installation.

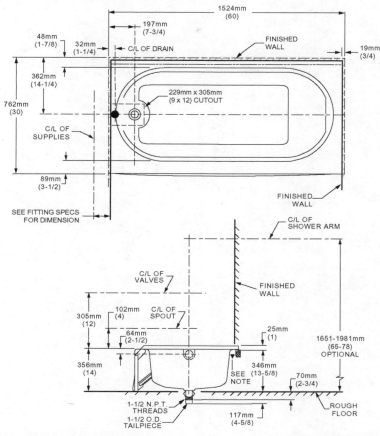

GENERAL SPECIFICATIONS FOR 2390/2391 BATHING POOL	
INSTALLED SIZE	60 x 30 x 14 In. (1524 x 762 x 356mm)
WEIGHT	110 Lbs. (50 Kg.)
WEIGHT w/WATER	460 Lbs. (209 Kg.)
GAL. TO OVERFLOW	42 Gal. (159 L)
BATHING WELL AT SUMP	42 x 19 In. (1067 x 483mm)
BATHING WELL AT RIM	56 x 25 In. (1423 x 635mm)
WATER DEPTH TO OVERFLOW	9-1/2 In. (241mm)
FLOOR LOADING	37 Lbs./Sq.Ft. (175 Kgs./Sq.m)
(PROJECTED AREA)	
PTS.	6.2
CUBE (FT³)	18.1

GENERAL SPECIFICATIONS FOR 2392/2393 BATHING POOL	
INSTALLED SIZE	60 x 30 x 17-1/2 In. (1524 x 762 x 445mm)
WEIGHT	119 Lbs. (54 Kg.)
WEIGHT w/WATER	469 Lbs. (213 Kg.)
GAL. TO OVERFLOW	42 Gal. (159 L)
BATHING WELL AT SUMP	42 x 19 In. (1067 x 483mm)
BATHING WELL AT RIM	56 x 25 In. (1423 x 635mm)
WATER DEPTH TO OVERFLOW	9-1/2 In. (241mm)
FLOOR LOADING	38 Lbs./Sq.Ft. (182 Kgs./Sq.m)
(PROJECTED AREA)	
PTS.	7.4
CUBE (FT³)	21.2

NOTES: LEFT HAND OUTLET SHOWN, RIGHT HAND REVERSE DIMENSIONS. (2391.202).

SHOWN WITH POP-UP C.D. & O.

REFER TO INSTALLATION INSTRUCTIONS SUPPLIED WITH FITTING.

CONCEALED PIPING NOT FURNISHED.

FITTINGS NOT INCLUDED AND MUST BE ORDERED SEPARATELY.

PROVIDE SUITABLE REINFORCEMENT FOR ALL WALL SUPPORTS.

▼ REFER TO INSTALLATION INSTRUCTIONS SUPPLIED WITH BATH.

IMPORTANT: Dimensions of fixtures are nominal and may vary within the range of tolerances established by ANSI Standard A112.19.4
These measurements are subject to change or cancellation. No responsibility is assumed for use of superseded or voided leaflet.

© 2003 American Standard Inc.

American Standard Inc.

Fitting Identification

Text Reference

Chapter 14, Pages 199–222

Objective

After completing this job, the student will be able to identify the various types of fittings and joints.

Instructions

1. Identify the cast-iron hub fittings.   Completed ☐

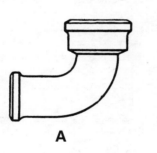

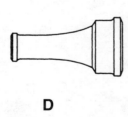

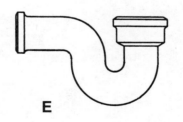

Cast Iron Soil Pipe Institute

_____ a. P-trap

_____ b. Y-branch

_____ c. Reducer

_____ d. 1/4 bend

_____ e. Sanitary tee

2. Identify the cast-iron no-hub fittings.   Completed ❑

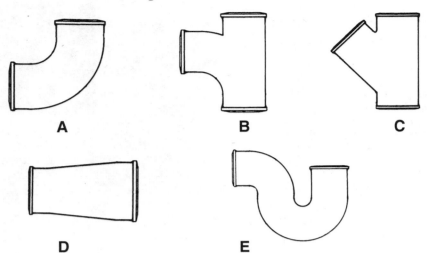

A   B   C

D   E

Cast Iron Soil Pipe Institute

_____ a. Y-branch

_____ b. Reducer

_____ c. P-trap

_____ d Sanitary tee

_____ e. 1/4 bend

3. Identify the parts of the lead and oakum joint.   Completed ❑

_____ a Hub

_____ b. Packed oakum

_____ c. Plain end/spigot bead

_____ d. 1″ deep lead

_____ e. Lead groove in hub

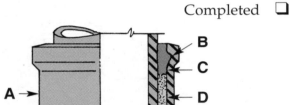

Cast Iron Soil Pipe Institute

4. Identify the parts of the compression joint.   Completed ❑

_____ a. Gasket

_____ b. Hub

_____ c. Lead groove

_____ d. Plain end

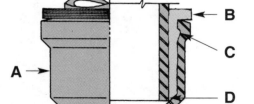

Cast Iron Soil Pipe Institute

Name _____

5. Identify the parts of the no-hub joint.

_____ a. Stainless steel shield

_____ b. Gasket

_____ c. Stainless steel retainer clamps

_____ d. No-hub pipe

Completed ❑

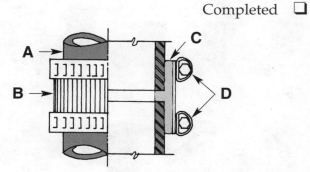

E.I. duPont de Nemours & Co.

6. Identify the P-trap fittings.

Completed ❑

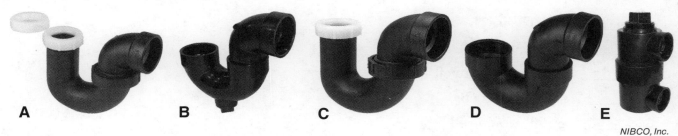

NIBCO, Inc.

_____ a. P-trap

_____ b. Swivel drum trap

_____ c. P-trap with slip joint

_____ d. P-trap with cleanout

_____ e. P-trap with union joint

7. Identify the DWV plastic fittings.   Completed ❑

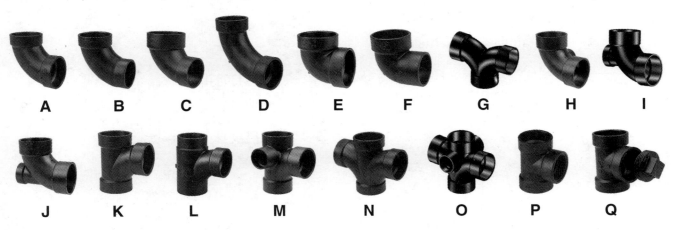

NIBCO, Inc.

_____ a. 90° ell

_____ b. 90° closet ell

_____ c. 90° long turn ell

_____ d. 90° street ell

_____ e. 90° double ell

_____ f. 90° street vent ell

_____ g. 90° ell with high heel inlet

_____ h. 90° ell with side inlet

_____ i. 90° vent ell

_____ j. 90° ell low heel inlet

_____ k. Double sanitary tee

_____ l. Sanitary tee

_____ m. Sanitary street tee

_____ n. Sanitary tee with 90° side inlet

_____ o. Test tee

_____ p. Vent tee

_____ q. Double sanitary tee/two 90° side inlets

Name _____

8. Identify the copper pressure fittings.   Completed ❑

A   **B**   **C**   **D**   **E**

F   **G**   **H**   **I**   **J**

NIBCO, Inc.

_____ a. Union

_____ b. FIPT adapter

_____ c. MIPT adapter

_____ d. Flush bushing

_____ e. Drop tee

_____ f. Tee

_____ g. 45° fitting ell

_____ h. 90° ell

_____ i. Reducing coupling

_____ j. Coupling

9. Identify the compression pressure fittings.   Completed ❑

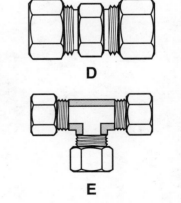

A   **B**   **C**   **D**

E

Parker Hannifin Corp.

_____ a. Compression union

_____ b. Compression ell

_____ c. Compression to MIPT adapter

_____ d. Compression tee

_____ e. Compression to MIPT 45° ell

10. Identify the malleable iron fittings.   Completed ❑

_____ a. Cap

_____ b. Bushing

_____ c. Reducer coupling

_____ d. Tee

_____ e. 45° street ell

_____ f. Pressure fitting

_____ g. Drainage fitting

A   **B**   **C**

D   **E**

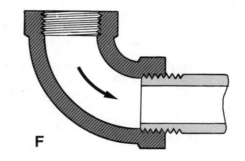

F

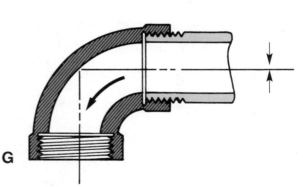

G

Goodheart-Willcox Publisher

Instructor's Initials _____

Date _____

Valve Identification

Text Reference

Chapter 15, Pages 226–232

Objective

After completing this job, the student will have demonstrated the ability to identify the parts of various valves.

Instructions

1. Identify the parts of the plastic ball valve.   Completed ❏

   _____ a. Handle

   _____ b. Handle clip

   _____ c. Stem

   _____ d. Body

   _____ e. Stem O-ring

   _____ f. Ball seat

   _____ g. Ball

   _____ h. Carrier

   _____ i. Carrier O-ring

   _____ j. End connector

   _____ k. Union nut

   _____ l. Face seal O-ring

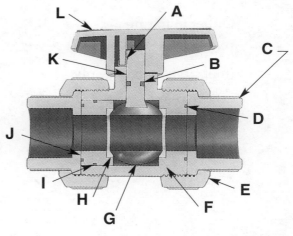

Celanese Plastics Co.

2. Identify the parts of the globe valve.   Completed ☐

_____ a. Inlet

_____ b. Outlet

_____ c. Valve seat

_____ d. Washer

_____ e. Handle

_____ f. Packing nut

_____ g. Packing box

_____ h. Bonnet

_____ i. Stem

_____ j. Screw thread

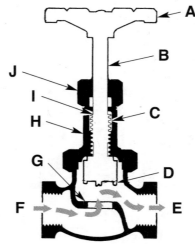

William Powell Co.

3. Identify the parts of the diaphragm flush valve.   Completed ☐

_____ a. Handle

_____ b. Auxiliary valve

_____ c. Upper chamber

_____ d. Diaphragm

_____ e. Bypass

_____ f. Outlet to fixture

_____ g. Plunger

_____ h. Inlet from water supply

_____ i. Lower chamber

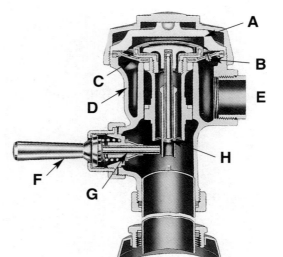

Sloan Valve Company

Name _____

4. Identify the parts of a swing check valve.   Completed ❑

      _____ a. Seat ring

      _____ b. Disc

      _____ c. Body

      _____ d. Disc nut

      _____ e. Cap

      _____ f. Hinge pin

The Fairbanks Co.

5. Identify the parts of a lift check valve.   Completed ❑

      _____ a. Disc

      _____ b. Body seat ring

      _____ c. Body

      _____ d. Cap

The Fairbanks Co.

Instructor's Initials _____

Date _____

DWV Pipe and Fitting Identification

Text Reference

Chapter 19, Pages 297–313

Objective

After completing this job, the student will be able to identify the fittings in a DWV pipe diagram.

Instructions

1. Identify the fittings in the DWV pipe diagram.

Completed ❑

Some letters may be used more than once.

_____ a. 2 x 2 x 2 wye

_____ b. 2 x 2 x 1-1/2 san. tee

_____ c. 2″ floor drain

_____ d. 2″ long radius 90° ell

_____ e. 1-1/2″ 45° ell

_____ f. 2″ 4° ell

_____ g. 2 x 1-1/2 x 1-1/2 san. tee

_____ h. 2 x 1-1/2″ bushing

_____ i. 2″ test tee with cleanout plug

_____ j. 1-1/2″ trap adapter

_____ k. 1-1/2″ 90° ell

_____ l. 1-1/2″ P-trap

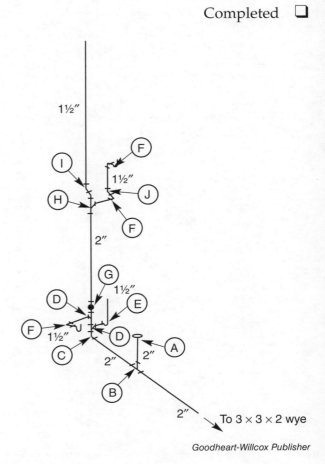

Goodheart-Willcox Publisher

2. Identify the fittings in the following DWV pipe diagram.

Completed ☐

_____ a. 3 x 3 x 3 test tee with cleanout

_____ b. 3 x 3 long sweep sanitary ell

_____ c. 4 x 3 closet flange

_____ d. 3 x 3 sanitary tee with 2-1/2″ side inlet

_____ e. 1-1/2 x 1-1/2 45° san. ell

_____ f. 3 x 3 90° san. ell

_____ g. 2 x 2 90° san. ell

_____ h. 3 x 3 x 1-1/2″ tee

_____ i. 1-1/2 x 1-1/2 90° ell

_____ j. 1-1/4 x 1-1/4 90° san. ell

_____ k. 1-1/4 x 1-1/4 45° san. ell

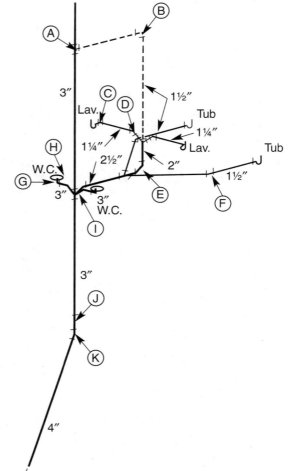

Instructor's Initials _____

Date _____

Job
9

DWV Fittings—No-Hub

Text Reference

Chapter 19, Pages 297–299

Objective

After completing this job, the student will be able to make a material list consisting of no-hub fittings and no-hub bands required for a simple bathroom.

Instructions

From the isometric diagram perform the following:

Completed ❑

1. Use the isometric diagram and list the no-hub fittings required for the system.

1½″ Sink

2″

1½″ Standpipe
1½″

Laundry tub

Floor drain
2″

2″

1½″
1¼″
Lav.
Tub
1¼″
2½″
2″ Lav.
Tub
1½″
W.C.
3″ 3″ W.C.

2½″

3″

4″

To Building Sewer

Goodheart-Willcox Publisher

2. Use the isometric diagram and list the no-hub bands required
for the system.   Completed ❏

Quantity	N0-Hub Fittings	No-Hub Bands	Size

Instructor's Initials _____

Date _____

Name _____ Date_____

Period_____ Course _____

Score_____

Job 10

Soldering Copper Pipe and Fittings

Introduction

Copper pipe and fittings may be joined by soldering. The connection is made watertight by using a filler metal known as solder. Heating the pipe and fittings with a plumber's torch melts the solder. The solder, which is now melted to a liquid state, is drawn into the socket of the fitting around the pipe and seals the connection. Although soldering is not difficult, the operation must be performed carefully for satisfactory results.

Text Reference

Chapter 9, Pages 127–130
Chapter 19, Pages 315–317

Tools and Equipment

Pipe vise
Tubing cutter
Measuring ruler or tape
Pencil for marking
Torch
Striker
Solder (lead free)
Soldering flux/paste brush
Sandpaper (emery cloth)
Cleaning cloth
3' of ¾" copper pipe
¾" 90° elbow
¾" copper male adapter
¾" copper cap
Safety goggles
Gloves

Objective

After completing this job, the student will have demonstrated the ability to join copper pipe and fittings using the soldering method. The connection must not leak when water pressure is applied. The instructor will determine the amount of pressure applied.

Warning
Follow all safety rules and guidelines that may apply to performing this activity.

Instructions

1. Place the copper pipe in the pipe vise and secure.   Completed ❏

2. Using a ruler, measure 12″ from the end of the pipe and mark   Completed ❏
   with a pencil.

3. Cut the pipe with the tubing cutter.   Completed ❏

Goodheart-Willcox Publisher

4. Ream each end to remove the burr. Remove the burr from the inside   Completed ❏
   of the tubing with a reamer.

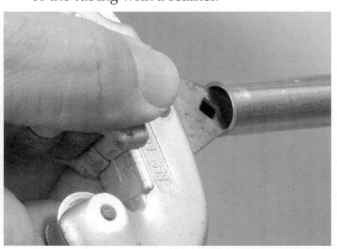

Goodheart-Willcox Publisher

Name _____

5. Clean the outside end of the pipe in the vise and one end of the 12″ cut pipe. Clean the outside of the tubing with sandpaper cloth or steel wool.

Completed ❑

6. Clean the inside of the sockets of the 90° elbow. Clean the inside of the fitting with a wire brush, steel wool, or sandpaper cloth.

Completed ❑

7. Immediately apply the proper amount of flux to all pipe and joint areas to be soldered. Apply flux to the outside of the tubing and the inside of the fitting.

Completed ❑

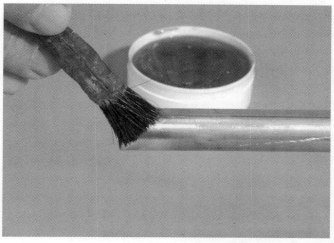

Goodheart-Willcox Publisher

8. Assemble the fluxed pipes into the fitting. Push and turn to ensure that the pipes are bottomed against the inside shoulders of the elbow fitting.

Completed ❑

9. Light the plumber's torch with the striker.

Completed ❑

Warning
Do not use a cigarette lighter or a match to light the torch. This could cause a severe burn injury.

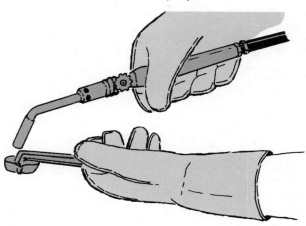

Goodheart-Willcox Publisher

10. Direct the heat onto the copper pipe and gradually move it toward the fitting. The inner cone of the flame should touch the metal. This is the hottest part of the flame.

Completed ❑

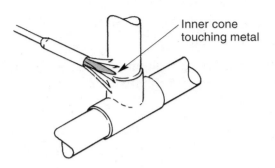

Inner cone touching metal

Goodheart-Willcox Publisher

11. Slowly touch the end of the solder to the joint area when the flux starts to boil. Apply solder and feed solder into the joint until there is a silver ring all around the fitting.

Completed ❑

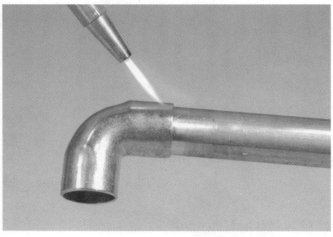

Goodheart-Willcox Publisher

12. Turn off the torch.

Completed ❑

13. Wipe the joint with a clean damp cloth while hot to remove excess solder and flux.

Completed ❑

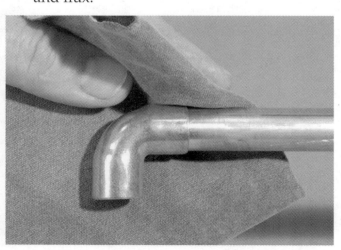

Goodheart-Willcox Publisher

Name _____

14. Once the pipe has cooled, solder the cap on the 12″ pipe side of the elbow and the copper adapter on the 24″ pipe side of the elbow. Follow the preceding steps under the direction of the instructor.

Completed ❏

15. Under the supervision of the instructor, test the piping connections with water pressure to ensure that there are no leaks.

Completed ❏

16. Clean up the work area and return all tools and material to the proper storage area.

Completed ❏

Instructor's Initials _____

Date _____

Joining Cast-Iron Hub and Spigot Type Pipe—Vertical Pipe

Introduction

Cast-iron soil pipe is used in nonpressure applications for building drainage systems. Some cast-iron soil pipe systems require the hub and spigot type of installation. When properly installed, these systems provide the leakproof connection necessary to prevent sewer gases from escaping after the system is in operation. The textbook covers several methods of testing a cast-iron soil pipe system.

Tools and Equipment

1—6" ruler
1—Plumber's lead melting furnace assembly, with burner, shield, and lead pot
1—Lead ladle
1—Yarning iron
1—Packing iron
1—Caulking irons set: inside and outside
1—Eight ounce ball peen hammer
1—Cast-iron soil pipe wheeler snapper chain cutter
1—Soapstone marker
1—Plumber's level (torpedo level)
1—Pair of safety goggles and gloves
1—Fire extinguisher
5'–4" single hub cast-iron pipe
Oakum (hemp treated with pitch to make it moisture proof)
Lead

Text Reference

Chapter 19, pages 320–321

Objective

After completing this job, the student will be able to join cast-iron hub and spigot type pipe, making a plumb, vertical joint. The student will be evaluated on safety and selection of the correct tools and materials. The student will also be graded on following the procedural steps, returning tools to their proper place, and cleaning the work area. This job should be completed within 45 minutes.

Note

Extra time may be allowed at the discretion of the instructor due to the hazards involved in this particular job. Think safety!

Instructions

1. Set up the lead melting furnace on a level surface in a well-ventilated area, and light it carefully. Be careful not to burn your hands.   Completed ❑

2. Take the 5′ sections of single hub pipe, measure from the spigot end, and mark with soapstone marker.   Completed ❑

3. Using a wheeler snapper chain cutter, cut the pipe on the mark.   Completed ❑

The Ridge Tool Co.

Warning

Wear safety goggles and take care not to pinch fingers between handles of chain cutters.

4. Insert cut piece of the pipe into the hub of the 4″ pipe.   Completed ❑

5. Place oakum into the joint by twisting it with one hand and forcing it down with the yarning iron in the other hand.   Completed ❑

6. After two circles of oakum are placed in the joint, drive the oakum down with the yarning iron and ball peen hammer.   Completed ❑

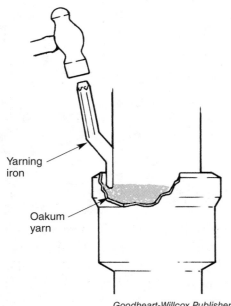

Yarning
iron

Oakum
yarn

Goodheart-Willcox Publisher

Name _____

7. Repeat step 6 until the oakum is 1" from the top of the hub.   Completed ❑

8. Using the torpedo level, straighten the pipe and hub until plumb.   Completed ❑

9. By this time the lead in the lead pot should be melted. Put on safety goggles and gloves, pick up ladle, and prepare to pour the lead joint.   Completed ❑

   Warning
   Allow ladle to warm up over furnace before submerging it into the hot lead because a cold or wet ladle will cause a violent reaction when submerged in the molten lead.

10. Submerge the ladle into the molten lead, dip the ladle full, and pour into hub.   Completed ❑

    Warning
    Use caution! The lead is hot.

11. When the lead cools to a solid, begin to caulk the joint using the inside caulking iron and striking the iron with the hammer on all four sides of the pipe to set the pipe and to keep it straight. Move the caulking iron the width of the iron until the inside of the joint is caulked.   Completed ❑

Caulking iron is used to compact lead into joint

Outside caulking iron

Inside caulking iron

Goodheart-Willcox Publisher

12. Using the outside caulking iron, caulk the outside of the hub using the same procedure as step 11.   Completed ❑

13. Secure the lead melting furnace, return tools and unused materials to the proper storage area, and clean the work area.   Completed ❑

14. Have the instructor check for plumbness and leaks.   Completed ❑

Instructor's Initials _____

Date _____

Job 12

Joining Cast-Iron No-Hub Pipe

Introduction

Cast-iron pipe is used primarily in nonpressure applications such as building drainage systems. Another term used for cast-iron pipe is *soil pipe*. Refer to the glossary in the textbook for further terminologies. The pipe and fittings are cast from gray iron. Cast-iron piping systems are selected because they are "quiet systems."

The purpose of this job is to join cast iron using the no-hub method, which allows for quicker installation. However, extra care must be taken to ensure proper support as specified by the Cast Iron Soil Pipe Institute. When properly installed, the no-hub system is trouble free.

Tools and Equipment

1—No-hub torque wrench
1—Stainless steel clamp with neoprene gasket
5"–2" No-hub cast-iron soil pipe
2" No-hub cast iron 1/8 bend fitting (45° elbow)

Text Reference

Chapter 19, Pages 318–320

Objective

After completing this job, the student will be able to join no-hub cast-iron pipe to a no-hub cast-iron fitting using a stainless steel clamp with a neoprene sleeve. The student will be evaluated for safety and the proper selection of tools and materials. The student will also be evaluated on following the procedural steps, the return of tools to their proper place, and cleanliness of the work area.

Note

The following procedure is the same in joining no-hub cast-iron pipe end to end.

Instructions

1. Take the neoprene sleeve out of the stainless steel clamp.   Completed ❑

Goodheart-Willcox Publisher

2. Slide the stainless steel clamp onto the cast-iron soil pipe.   Completed ❑
3. Put the end of pipe and fitting into the neoprene gasket.   Completed ❑

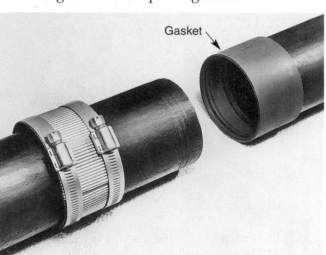

Gasket

Goodheart-Willcox Publisher

Caution

Be sure that the pipe and fitting fit to the center of the gasket since this will cause the fitting to be proper. Failure to perform step 3 correctly will cause the fitting to be loose (incorrect connection) and leak.

Name _____

4. Slide the stainless steel clamp onto the neoprene gasket and tighten the clamp evenly with the torque wrench.

Completed ❑

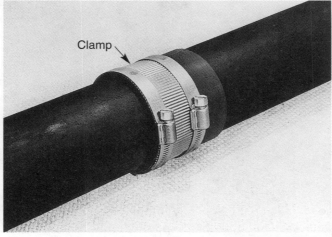

Goodheart-Willcox Publisher

5. Have the instructor check the connection to make sure the clamp is evenly tightened, square, and watertight at atmosphere pressure.

Completed ❑

Instructor's Initials _____

Date _____

Job 13

Joining PVC Pipe

Introduction

The use of polyvinyl chloride (PVC) pipe for plumbing is increasing rapidly due to low material cost and ease of installation. One advantage is that fewer hand tools are required. PVC is used in pressure (water supply) and nonpressure (DWV) applications. The technique of joining by solvent cement is relatively simple. When done correctly, it produces a leak-free connection. The student must always read the directions on the solvent cement container. The directions provide information regarding the hazards of using the cement and cleaner fluids.

Note

There is no such thing as "all-purpose cement". The strength of the cement is applicable to the pipe size. Therefore, refer to the manufacturer's specifications and directions. Remember, using the wrong strength cement can result in a leaking joint. Also, dissimilar plastics are not compatible to join together.

Tools and Equipment

1—6' ruler
1—Fine tooth handsaw
1—Miter box
1—Half round file
1—Marking pencil
1—2" PVC pipe (longer than 24")
1—2" PVC 90° elbow
1—Wiping cloth
PVC primer and brush
PVC solvent cement and brush
Safety goggles

NOTE

Be sure the proper solvent is used because there is no universal solvent cement.

Text Reference

Chapter 19, Pages 313–315

Objective

After completing this job, the student will be able to join PVC pipe and fittings using the proper primer and solvent cement. The student will be evaluated on the safety and selection of the correct tools, materials, and solvent cement. The student will also be evaluated on the return of tools and materials, and the cleanliness of the work area.

Instructions

1. Using the 6′ ruler, mark 24″ on the length of 2″ PVC pipe.   Completed ❏

2. Place the pipe in the miter box and cut the pipe with the fine tooth saw.   Completed ❏
   Be sure that the cut is square with no more than plus or minus 1/8″.

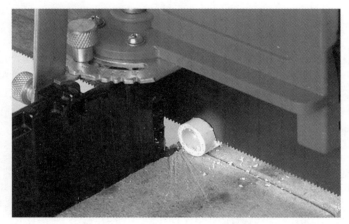

Goodheart-Willcox Publisher

3. Remove burrs from the inside of the pipe using the half round file.   Completed ❏

4. Remove any ridges on the outer edges of pipe cut.   Completed ❏

5. Use the wiping cloth to remove any dirt and grit from the pipe end and   Completed ❏
   from the inside of the 2″ PVC elbow sockets.

6. Using the primer application brush, apply an even coat of primer to the   Completed ❏
   end of the pipe as far as it will be inserted into the fitting. Apply an even
   coat of primer in the fitting socket.

   Warning
   Take care while using the primer and solvent cement because the fumes from both
   are extremely dangerous and highly flammable. Wear safety goggles.

7. Wait 15 seconds, then apply an even coat of solvent cement to the end of   Completed ❏
   the pipe to the depth that will be inserted into the fitting.

Name _____

8. Apply an even coat of solvent cement to the inside surface of the fitting.   Completed ❏

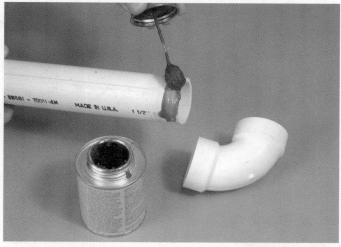

Goodheart-Willcox Publisher

9. Immediately after applying the solvent cement, insert the pipe into the   Completed ❏
    fitting, twisting the pipe one-quarter of a turn until it bottoms in the
    socket. Make the twist in one direction only. Hold the joint firmly in
    place for about 10 seconds (longer in cooler weather) to allow the two
    surfaces to start bonding together.

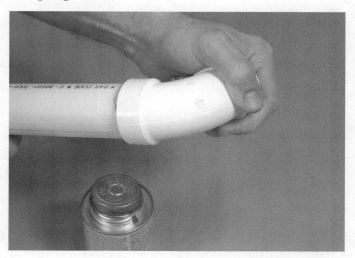

Goodheart-Willcox Publisher

10. Check the ring of solvent cement to see if it has been pushed out all the way around the joint during assembly and alignment. If the ring does not go all the way around the joint, you have not used enough solvent cement and the joint could leak.

Completed ❑

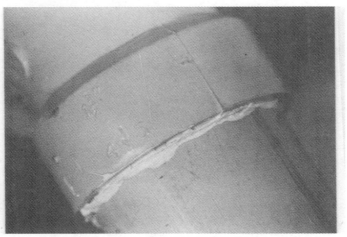

Goodheart-Willcox Publisher

11. Wipe off the excess solvent if the ring looks complete. Removing the excess solvent helps the joint cure faster.

Completed ❑

12. Return tools and materials to their proper places and clean the work area.

Completed ❑

13. Have the instructor check the joint for proper measurement, proper alignment, and leaks.

Completed ❑

Instructor's Initials _____

Date _____

Job
14

Joining Galvanized Steel Pipe and Sleeve Coupling

Introduction

Galvanized steel pipe is used for hot and cold water distribution, steam and hot water heating, gas and air piping, and drainage and vent piping. One method of joining galvanized steel pipe is by threading the pipe and then using it with threaded fittings. The threads are tapered to form a watertight joint when tightened securely.

Tools and Equipment

2—14″ pipe wrenches
1—Yoke or chain vise
1—3/4″ ID galvanized sleeve coupling
1—3/4″ galvanized cap
2—3/4″ x 6″ galvanized nipples
Pipe joint sealer or Teflon™ tape

Text Reference

Chapter 19, Pages 317–318

Objective

When provided with the proper tools and materials, the student will be able to join steel pipe with a sleeve coupling, producing a joint that will not leak when pressure is applied. The student will be evaluated by the instructor on the procedural steps of this job and will also be graded on the selection of tools, the return of the tools selected, and the cleanliness of the work area.

Note
The following procedure will be the same whether the pipe used is galvanized, black steel, or wrought iron.

Instructions

1. Secure one 3/4″ x 6″ nipple in the yoke or chain vise.   Completed ❑

   Caution
   Do not tighten the vise on threads. This will damage the threads and cause the connection to leak.

2. Apply pipe joint sealer or Teflon™ tape to perfect the threads on the nipple in the vise.   Completed ❑

Goodheart-Willcox Publisher

3. Screw the sleeve coupling onto the pipe threads and turn it clockwise until it is hand tight.   Completed ❑

4. Place the wrench on the sleeve coupling and turn it clockwise one to three turns until firmly tight.   Completed ❑

Goodheart-Willcox Publisher

   Caution
   Do not overtighten. Overtightening will spread the sleeve coupling and cause the joint to leak.

5. Apply pipe joint sealer or Teflon™ tape to one end of the other 3/4″ x 6″ nipple and turn clockwise in sleeve coupling until hand tight.   Completed ❑

6. Place the pipe wrench on the second nipple and turn it clockwise one to three turns until firmly tight.   Completed ❑

   Caution
   Do not overtighten because this will spread the sleeve coupling and cause the joint to leak.

Name _____

7. Apply pipe joint sealer or Teflon™ tape to perfect the threads on the nipple.   Completed ❑

8. Screw the 3/4″ cap on threads of the nipple and turn clockwise until hand tight.   Completed ❑

9. Place the pipe wrench on the cap and turn it clockwise one to three turns until firmly tight.   Completed ❑

10. Have the instructor check for leaks by applying water pressure.   Completed ❑

Instructor's Initials _____

Date _____

Job
15

Cutting and Threading Galvanized Steel Pipe

Introduction

Galvanized steel pipe is used for hot and cold water distribution, steam and hot water heating, gas and air piping, and drainage and vent piping. One method of joining galvanized steel pipe is by threading the pipe, which may then be used with threaded fittings. The threads are tapered to form a watertight joint when tightened securely.

Tools and Equipment

1—Yoke or chain vise
1—6' ruler
1—Marking pencil
1—Steel pipe cutter
1—Steel pipe reamer
1—1/2" ratchet pipe threader
21'-1/2" ID galvanized steel pipe
1—Wire brush
Oiler
Cutting oil
Safety goggles

Text Reference

Chapter 19, Pages 317–318

Objective

When provided with the proper tools and materials, the student will be able to thread galvanized steel pipe using the ratchet threader. The student will be evaluated by the instructor on the procedural steps of this activity and will cut pipe within plus or minus 1/8" of the measurement provided. The student will also be graded on the selection and return of tools and cleanliness of the work area.

Instructions

1. Secure the pipe in the yoke or chain-type vise.   Completed ❑

Pipe Vise (Yoke)

Chain-Type Vise

The Ridge Tool Co.

2. Using a 6′ ruler, measure 12″ from the end of the galvanized steel pipe   Completed ❑
   and mark it with the marking pencil.

3. Open the steel pipe cutters by turning the T handle counterclockwise   Completed ❑
   until the cutter slips over the 1/2″ galvanized pipe. Then turn the
   T handle clockwise to tighten the pipe cutter around pipe on the mark
   that will line up with the cutter wheel. Snug the cutter around pipe and
   make one revolution to see if the cutter wheel is tracking properly.

   Warning
   Wear safety goggles.

Name _____

4. If the cutter is tracking properly, continue making revolutions around the pipe and tighten the cutter 1/4 turn for each revolution until the pipe separates.

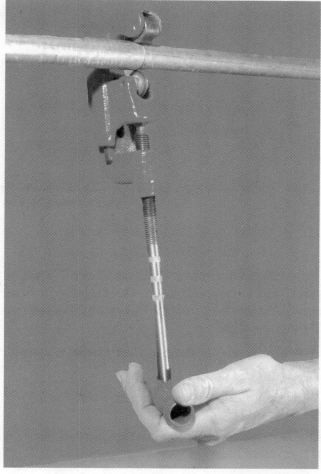

The Ridge Tool Co.

Caution
Do not overtighten. Overtightening will damage/break the cutter wheel or cause an excessive burr to form inside the pipe.

5. Ream the inside of the pipe to remove the burr caused by the pipe cutter   Completed ❏
   and restore the inside diameter to its original size.

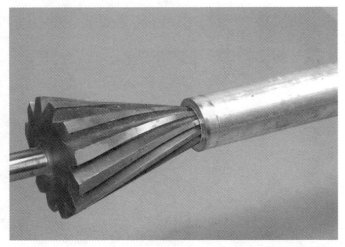

Goodheart-Willcox Publisher

Caution
Do not over ream the pipe, since this will make the pipe end weak.

6. Using the 1/2″ ratchet die threader, press the die head against the pipe   Completed ❏
   with one hand, and then turn the ratchet handle clockwise with the other
   hand. Apply pressure until the die teeth catch. Lubricate the dies with
   cutting oil as they feed onto the pipe.

7. When the dies are flush with the end of the pipe, the threads are at   Completed ❏
   standard length.

Caution
Use plenty of cutting oil to preserve the pipe dies.

8. Remove die threaders by reversing lock pin and turning ratchet handle   Completed ❏
   counterclockwise until threaders are removed from pipe.

9. Use a wire brush to clean the threads on the end of the pipe.   Completed ❏

Warning
Pipe threads are very sharp and can cut the hands and fingers if you are not careful.

10. Have the instructor check for proper threads and measurement of pipe to   Completed ❏
    plus or minus 1/8″.

11. Clean the work area and replace all tools and materials to their proper   Completed ❏
    place of storage.

Instructor's Initials _____

Date _____

Job
16

Water Supply

Text Reference

Chapter 20, Pages 323–327

Objective

After completing this job, the student will be able to make a copper or PVC/CPVC fitting material list for a water supply system.

Instructions

1. From the isometric diagram of the water supply system, make a copper   Completed ❏
   or PVC fitting material list (take off list).

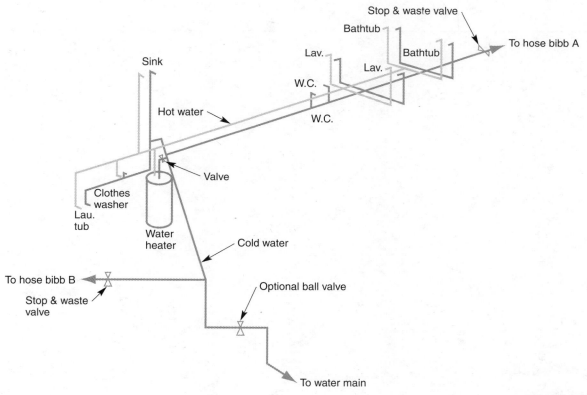

Goodheart-Willcox Publisher

Quantity	Material	Size	Completed

Instructor's Initials _____

Date _____

Name _____ Date_____

Period_____ Course _____

Score_____

Joining Copper Pipe Using Compression Fittings

Introduction

Copper pipe may be joined by compression fittings. The connection is made watertight by the compression of a brass ferrule around the copper pipe as the nut is tightened on the fitting.

Text Reference

Chapter 22, Pages 375–377

Tools and Equipment

1—6" crescent wrench
1—8" crescent wrench
1—Tubing cutter
1—Ruler
1—Pencil or marking device
1—3/8" OD compression type fitting
12" x 3/8" OD copper tubing

Note

Be sure the compression fitting has a brass ferrule.

Objective

After completing this job, the student will be able to join copper tubing using compression connectors. The connections will not leak when water pressure is applied.

Instructions

1. Using a ruler, measure 6" from one end of the 3/8" OD copper tubing   Completed ❏
   and mark with a pencil.

2. Cut the pipe on the mark and ream.

Goodheart-Willcox Publisher

3. Remove the compression nut and ferrule from one side of the   Completed ❑
   compression fitting.

4. Slip the compression fitting nut over the end of the pipe, then slip   Completed ❑
   the ferrule over the end of the pipe. Now insert the pipe end into the
   compression fitting socket.

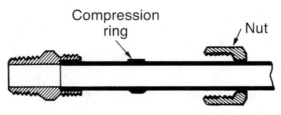

Compression
ring   Nut

Goodheart-Willcox Publisher

5. Slide the ferrule and nut up to the compression fitting and hand tighten.   Completed ❑

6. Hold the compression fitting with the 8″ crescent wrench and tighten the   Completed ❑
   compression nut with the 6″ crescent wrench until firmly tight.

7. Return tools and materials to their proper storage area. Clean the work   Completed ❑
   area.

Instructor's Initials _____

Date _____

Flaring and Connecting Copper Tubing

Introduction

Copper pipe may be joined by using a flare type connection. This connection is made by flaring the tubing at a 45° angle, which fits to a beveled connector using a flare nut.

Text Reference

Chapter 22, Pages 375–377

Tools and Equipment

1—Rigid #10 tubing cutters
1—6' ruler
1—Flaring tool
1—Marking pencil
1—Reaming tool
1—Roll of 1/2" OD copper tubing
1—1/2" OD flare nut
1—1/2" MIP x 1/2" flare adaptor

Objective

After completing this job, the student will be able to cut, ream, and flare 1/2" copper tubing and make a flare connection that will not leak when pressure is applied.

Instructions

1. Using 6' ruler, measure 8" from the end of the roll of copper tubing and mark with the marking pencil. Completed ❑

2. Open tubing cutter to fit over tubing and place cutter wheel on mark. Completed ❑

3. Tighten handle on cutter by turning it clockwise until it fits snug on the tubing.

Completed ❏

Goodheart-Willcox Publisher

4. Turn the tubing cutters around the pipe, tightening the handle 1/4 turn every three revolutions.

Completed ❏

Caution
Do not overtighten. Overtightening will damage the cutter wheel.

5. Turn and tighten repeatedly until tubing is cut and separated.

Completed ❏

6. Ream the burrs out of the tubing, using the reamer located in the tubing cutter.

Completed ❏

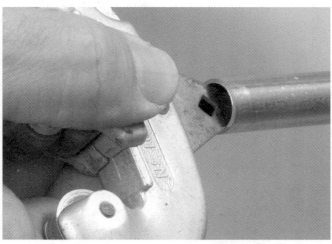

Goodheart-Willcox Publisher

Name _____

Caution

Ream only down to the actual diameter of the pipe. Over reaming will cause the pipe walls to become thin and result in cracking and splitting while attempting to flare tubing.

7. Slide the flare nut over the pipe with the threads in the direction of the end of the pipe that is to be flared.

Completed ❏

8. Insert the tubing end into the flaring tool forming block that is the same as the pipe. Allow the end of the pipe to be flush with the flaring block.

Completed ❏

9. Tighten the blocks together.

Completed ❏

Imperial Eastman Corp.

10. Slide the flare handle over the block until it centers over the tubing.

Completed ❏

11. Lock the flare handle in place by slowly tightening the handle in the clockwise direction.

Completed ❏

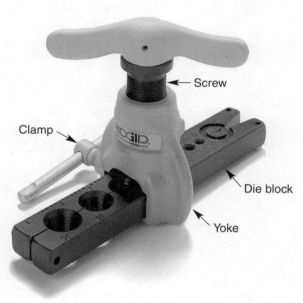

The Ridge Tool Co.

12. Tighten and flare the tubing until the handle becomes snug.   Completed ❑

 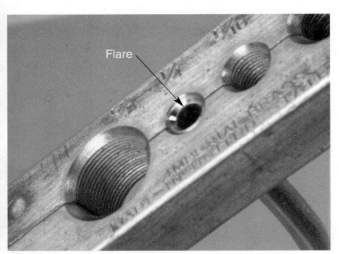

Flare

Goodheart-Willcox Publisher

Caution
Do not overtighten. Overtightening will crack and split the tubing.

13. Remove the flare tool by turning the handle counterclockwise.   Completed ❑
14. Remove the flare block by loosening the blocks.   Completed ❑
15. Check the flare for possible cracks or splits.   Completed ❑
16. Slide the flare nut over the flare end of the tubing and tighten.   Completed ❑
17. Check for leaks.   Completed ❑
18. Return tools and equipment to their proper storage area. Clean the work area.   Completed ❑

Instructor's Initials _____

Date _____

Faucet and Rotating Ball Faucet Identification

Text Reference

Chapter 29, Pages 452–458

Objective

After completing this job, the student will have demonstrated the ability to identify the parts of a compression faucet and rotating ball faucet.

Instructions

1. Identify the parts of the faucet.   Completed ☐

   _____ a. Handle

   _____ b. Packing

   _____ c. Spline

   _____ d. Bonnet

   _____ e. Stem

   _____ f. Spout

   _____ g. Internal thread

   _____ h. Faucet base

   _____ i. Seat

   _____ j. Brass screw

   _____ k. Washer

   _____ l. Stem threads

Goodheart-Willcox Publisher

2. Identify the parts of the faucet.   Completed ❑

_____ a. Cam housing

_____ b. Aerator

_____ c. Spout

_____ d. Cap

_____ e. Setscrew

_____ f. Ball assembly

_____ g. Cam rubber

_____ h. Diverter unit

_____ i. Stainless steel spring

_____ j. Seat washer

_____ k. O-ring washers

_____ l. O-ring

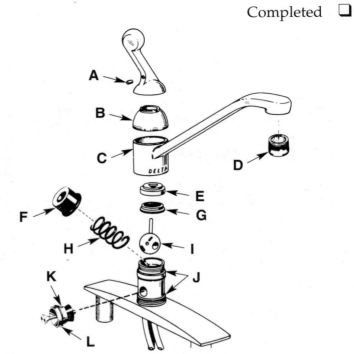

Goodheart-Willcox Publisher

Instructor's Initials _____

Date _____

Job 20

Installing a Toilet (Tank-Type)

Introduction

Toilets are either wall-hung or floor-mounted. The wall-hung units are often installed in commercial buildings to make floor cleaning easier. Toilets are available in a variety of styles, so it is necessary to refer to the manufacturer's instructions before beginning installation. The installation procedures discussed in this activity apply to most designs. Most codes require a valve at the toilet stub-out to allow shutting off the water supply to service the toilet. The tools and supplies required for installation of the stub-out valve depend on the type of pipe and fittings used for the cold water piping.

Text Reference

Chapter 22, Pages 370–373

Tools and Equipment

8" Crescent wrench
10" Crescent wrench
Channel lock pliers
Tubing cutter
Level
Pipe joint compound
Toilet bowl wax ring
Angle shut off valve, wall flange, toilet tank supply
Closet bolts, washers, nuts
Closet flange
Toilet unit (bowl and tank)
Toilet seat

Objective

After completing this job, the student will have demonstrated the ability to install a tank-type toilet. The toilet will flush and function properly without any leaks.

Instructions

1. Turn off water supply serving the fixture branch.   Completed ❏

2. Cut off the water stub-out.   Completed ❏

3. Install the angle type shutoff valve.   Completed ❏

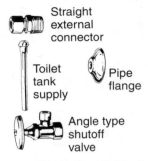

Chicago Specialty Mfg., Co.

4. Install the closet flange to the DWV piping system and flange bolts.   Completed ❏
    Adjust the flange so the slots are equally centered from the back behind
    the toilet.

    Note
    The bottom of the flange should be flush with the finished floor.

5. To install the bowl, refer to the following exploded view and proceed as   Completed ❏
    follows:

    A. Place the bowl temporarily over the
        closet flange. Check for levelness. If it is
        not level, insert shims to level the bowl.

    B. Lift the bowl off the flange and turn it
        upside down. Fit the wax seal ring onto
        the discharge opening.

    C. Position the bowl carefully over the
        closet flange and closet bolts. Check
        again to make sure the bowl is level and
        squarely seated. Then, place washers on
        the closet bolts and install the closet
        nuts on the bolts. Tighten nuts securely
        and place bolt covers over bolts.

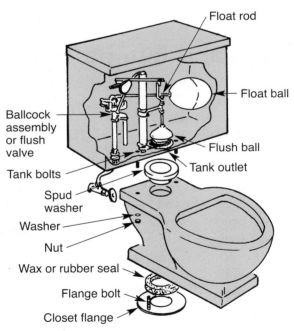

Goodheart-Willcox Publisher

Name _____

6. To install the tank, review the instructions provided with the toilet and again refer to the exploded view. The ballcock, flush valve, and flush valve lever may already be assembled. It is a good idea to check to make sure all connections are tight. To install the tank and fittings:

Completed ❑

A. Secure the ballcock assembly to the bottom of the tank making sure the gasket provides a complete seal around the opening.

B. Place a spud washer over the water inlet hole of the bowl.

C. Carefully place the tank in position so that its opening fits over the spud washer. Press the tank into place.

D. Secure with tank bolts.

E. Install the float rod and float ball on flush valve if necessary.

F. Attach the flush lever to the flush valve.

G. Connect the water supply using a flexible water supply tube or flexible chrome plated copper toilet tank supply tube.

H. Once the connections are secured, turn on the water supply, open the angle stop valve, and allow the tank to fill with water. Check all connections for leaks.

I. Once any leaks have been corrected, flush the toilet to check for leaks at the toilet bowl seal.

J. Install the toilet seat, making sure it is properly aligned over the bowl.

K. Have the instructor check the installation.

Instructor's Initials _____

Date _____

Rough-In Residential Bathroom Group

Introduction

Roughing in a residential bathroom group provides an opportunity for students to demonstrate skills that have been learned in the plumbing class and lab. Basic plumbing installation skills will be used for installing a DWV piping system using PVC pipe and fittings and for installing a water piping system using copper pipe and fittings. Students will refer to the fixture manufacturer's specification instructions before beginning installation. Students will be required to rough-in the water supply and the drainage and vent system for a three-piece bath using PVC and copper pipe and fittings. The sole plate should be cut out for the stacks before installation is begun.

Tools and Equipment

4′ x 8′ bathroom mock-up with a 2″ x 6″ stud wall
6′ folding rule or steel tape
Pencil or marking pen
Copper tubing cutter and reamer
PVC saw
PVC reamer
Battery powered drill–3/4″ or larger wood bit
Claw hammer
Keyhole saw
Wood screws
Standard screwdriver
Soldering paste
Copper cleaning brushes/emery cloth
Channel lock pliers
Soldering torch/tip
Small gas tank cylinder
Lead-free solder
PVC glue and cleaner
Hard hat
Safety goggles
Clean wiping rag
Fire extinguisher
Test plugs (3″, two–1-1/2″)
Riser clamps (two–1-1/2″)

Objective

After completing this job, the student will have demonstrated the ability to rough-in the DWV piping system and the water supply piping system for a three-piece bathroom group.

Instructions

1. Read the requirements for the DWV piping system and the water supply piping system before beginning the installation.

   Caution

   Comply with all safety rules and procedures.

   Completed ❑

2. Perform DWV system installation requirements.

   Completed ❑

   A. Soil and waste lines must be installed to a grade of not less than 1/8″ fall per foot. (Wooden blocks of 1/2″, 3/4″ and 1″ can be used to ensure proper grade.)

   B. Vertical pipes must be plumb.

   C. Horizontal vent lines must slope from the main stack back to the source.

   D. The 3″ vent stack must extend 12″ above the top plate to the center of the 3 x 2 vent tee.

   E. The top of the platform is considered finished floor unless directed otherwise.

   F. The center line of fixtures shall match the floor plan and the manufacturer's specification sheets, plus or minus 1/4″.

   G. The closet flange shall be attached to the platform with wood screws.

   H. Riser clamps will be installed at the floor level of each stack.

   I. The DWV drainage system shall be water tested to check for any leaks.

3. Perform cold and hot water system installation requirements.

   Completed ❑

   A. The cold and hot water supply system shall be installed to comply with the isometric drawing shown on Drawing #3 and shall extend 8″ beyond the platform wall on the tub end and be tied back together as shown.

   B. Cold and hot water lines for the bathtub shall extend up to the height of the tub/shower valve and be looped together, as shown in Drawing #3.

   C. The water supplies for lavatory, toilet, and tub/shower shall be installed in accordance with the manufacturer's specification sheets, plus or minus 1/8″.

   D. Horizontal lines must be level, and vertical lines must be plumb.

   E. Fixture supply stub-outs shall extend a minimum of 5″ beyond the wall.

   F. The water supply system shall be water tested.

Name _____

4. Upon completion, clean the work area and return tools to storage.   Completed ☐

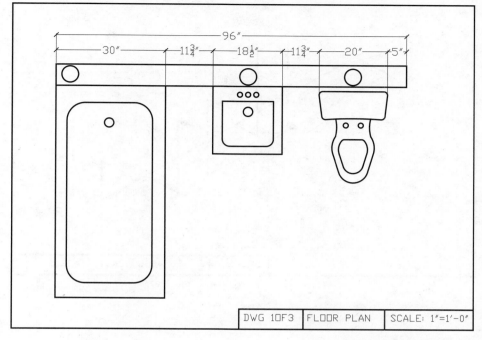

Goodheart-Willcox Publisher

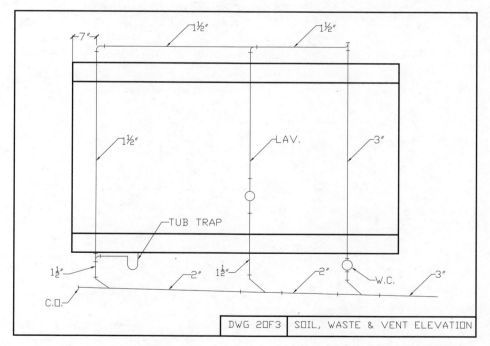

Goodheart-Willcox Publisher

Instructor's Initials _____

Date _____

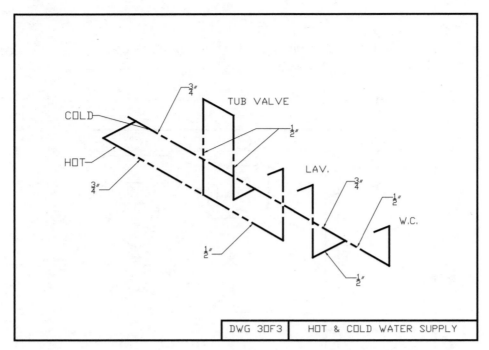

COLD

HOT

TUB VALVE

LAV.

W.C.

DWG 30F3   HOT & COLD WATER SUPPLY

American Standard

DECLYN™
WALL-HUNG LAVATORY
VITREOUS CHINA

DECLYN WALL-HUNG LAVATORY

- Vitreous china
- Rear overflow
- Soap depression
- Faucet ledge.
    Shown with 2000.101 Ceramix faucet
    (not included)

❏ **0321.026** With wall hanger (Illustrated)
Faucet holes on 102mm (4") centers

❏ **0321.075** For concealed arms support
Faucet holes on 102mm (4") centers

Nominal Dimensions:
483 x 432mm
(19" x 17")

Bowl sizes:
362mm (14-1/4") wide,
273mm (10-3/4") front to back,
152mm (6") deep

Fixture Dimensions conform to ANSI Standard
A112.19.2

To Be Specified	
❏	Color:
❏	Faucet*:
❏	Faucet Finish:
❏	Supplies:
❏	1-1/4" Trap:
❏	Nipple:
❏	Concealed Arms Support (by others):

* See faucet section for additional models available

● Top of front rim mounted 864mm (34") maximum from
finished floor.
**MEETS THE AMERICAN DISABILITIES ACT GUIDELINES
AND ANSI A117.1 REQUIREMENTS FOR PEOPLE WITH
DISABILITIES**

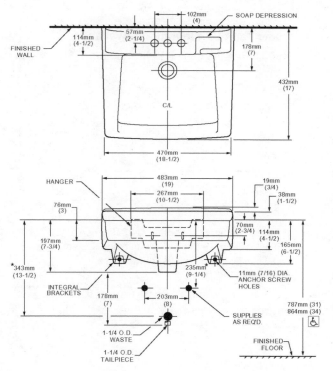

NOTES:
★ DIMENSIONS SHOWN FOR LOCATION OF SUPPLIES AND "P" TRAP ARE SUGGESTED.
PROVIDE SUITABLE REINFORCEMENT FOR ALL WALL SUPPORTS.
FITTINGS NOT INCLUDED AND MUST BE ORDERED SEPARATELY.
IMPORTANT: Dimensions of fixtures are nominal and may vary within the range of tolerances
established by ANSI Standard A112.19.2.
These measurements are subject to change or cancellation. No responsibility is assumed for use
of superseded or voided pages.

SPS 0321

© 1995 American Standard Inc.

Revised 4/97

American Standard Inc.

American Standard

COLONY™ ELONGATED TOILET
VITREOUS CHINA

COLONY™ ELONGATED TOILET

❑ **2399.012**
- Vitreous china
- Low-consumption (6.0 Lpf/1.6 gpf)
- Elongated siphon action jetted bowl
- Fully glazed 2" trapway
- American Standard Aquameter™ Water Control
- Large 10" x 8" water surface area
- Color-matched trip lever
- Sanitary bar on bowl
- 2 bolt caps
- 100% factory flush tested

❑ **3344.017** Bowl

❑ **4392.016** Tank

Nominal Dimensions:
762 x 508 x 743mm (30" x 20" x 29-1/4")

Fixture only, seat and supply by others

Alternate Tank Configurations Available:

❑ **4392.500** Tank complete w/Aquaguard Liner

❑ **4392.800** Tank complete w/Trip Lever Located on Right Side

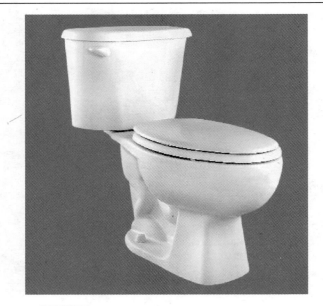

To Be Specified

❑ Color:

❑ Seat: American Standard #5324.019 "Rise and Shine" (with easy to clean lift-off hinge system) solid plastic closed front seat with cover. See pageTB-001.

❑ Seat: American Standard #5311.012 "Laurel" molded closed front seat with cover. See pageTB-001.

❑ Alternate Seat:

❑ Supply with stop:

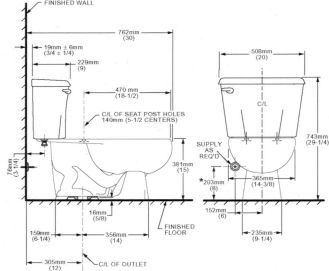

NOTES:
THIS COMBINATION IS DESIGNED TO ROUGH-IN AT A MINIMUM DIMENSION OF 305MM (12") FROM FINISHED WALL TO C/L OF OUTLET.
★ DIMENSION SHOWN FOR LOCATION OF SUPPLY IS SUGGESTED.
SUPPLY NOT INCLUDED WITH FIXTURE AND MUST BE ORDERED SEPARATELY.
IMPORTANT: Dimensions of fixtures are nominal and may vary within the range of tolerance established by ANSI Standard A112.19.2
These measurements are subject to change or cancellation. No responsibility is assumed for use of superseded or voided pages.

Compliance Certifications -
Meets or Exceeds the Following Specifications:
- ASME A112.19.2M (and 19.6M) for Vitreous China Fixtures - includes Flush Performance, Ball Pass Diameter, Trap Seal Depth and all Dimensions

SPS 2399.012

© 2000 American Standard Inc.

American Standard Inc.

American Standard

PRINCETON™ RECESS BATH
AMERICAST® BRAND ENGINEERED MATERIAL

PRINCETON RECESS BATH

Americast® brand engineered material

- ❏ **2391.202** Right Hand Outlet
- ❏ **2391.202TC** Same as above w/tub cover
- ❏ **2390.202** Left Hand Outlet
- ❏ **2390.202TC** Same as above w/tub cover
- • Acid resistant porcelain finish
- • Recess bath with integral apron and tiling flange
- • Integral lumbar support
- • Beveled headrest
- • Full slip-resistant coverage
- • End drain outlet

PRINCETON RECESS BATH for Above Floor Rough Installation

- ❏ **2392.202** Left Hand Outlet for above floor installation
- ❏ **2392.202TC** Same as above w/tub cover
- ❏ **2393.202** Right Hand Outlet for above floor installation
- ❏ **2393.202TC** Same as above w/tub cover

Nominal Dimensions: 1524 x 762 x 356mm (445mm for above floor installation)
60" x 30" x 14" (17-1/2" for above floor installation)

Bathing Well Dimensions: 1423 x 635 x 337mm (56" x 25" x 13-1/4")

Americast® brand engineered material is a composition of porcelain bonded to enameling grade metal, bonded to a patented structural composite.

Compliance Certifications -
Meets or Exceeds the Following Specifications:
- • ASME A112.19.4 for Americast Plumbing Fixtures
- • ASTM F-462 for Slip-resistant Bathing Facilities
- • ANSI Z124.1 Ignition Test
- • ASTM E162 for Flammability
- • NFPA 258 for Smoke Density

NOTE: Roughing-in dimensions shown on reverse side of page.

To Be Specified
❏   Color:
❏   Bath Faucet*:
❏   Faucet Finish:
❏   Drain:
❏   Drain Finish:

* See faucet section for additional models available

BV-38

Revised 5/98   © 2003 American Standard Inc.

American Standard Inc.